牛卵泡可卡因-苯丙胺调节转录肽（CART）受体的筛选

Screening of Cocaine-and Amphetamine-Regulated Transcript Peptide (CART) Receptor of Cattle Follicle

李鹏飞 著

中国农业科学技术出版社

图书在版编目（CIP）数据

牛卵泡可卡因－苯丙胺调节转录肽（CART）受体的筛选／李鹏飞著．—北京：中国农业科学技术出版社，2015.10
ISBN 978－7－5116－2330－0

Ⅰ.①牛… Ⅱ.①李… Ⅲ.①牛－丘脑－肽－研究 Ⅳ.①S852.1

中国版本图书馆 CIP 数据核字（2015）第 252757 号

责任编辑	张孝安
责任校对	贾海霞

出 版 者	中国农业科学技术出版社
	北京市中关村南大街 12 号　邮编：100081
电　　话	（010）82109708（编辑室）　（010）82106624（发行部）
	（010）82109703（读者服务部）
传　　真	（010）82106650
网　　址	http://www.castp.cn
经 销 者	各地新华书店
印 刷 者	北京富泰印刷有限责任公司
开　　本	710 mm×1 000 mm　1/16
印　　张	10
字　　数	150 千字
版　　次	2015 年 10 月第 1 版　2016 年 1 月第 2 次印刷
定　　价	36.00 元

◁ 版权所有·翻印必究 ▷

国家自然科学基金项目（No.31172211）；农业部"948"项目（No.2010 - Z43）；国家自然科学基金项目（No.31402156）；山西省横向协作与委托项目（2010HX54）；山西省回国留学人员科研资助项目（2014 - 重点 5）；山西省科技攻关项目（No.20130311027 -2）；山西省人事厅人才引进项目；山西省青年基金（No.2014021028 -2）；山西农业大学引进人才博士科研启动费（2014ZZ04）；山西农业大学青年拔尖创新人才支持计划（TYIT201403）和科研管理费资助重大项目和标志性成果培育项目（71060003）部分研究成果。

内容简介
ABSTRACT

  CART 是一种下丘脑分泌的神经肽，在动物体内具有广泛的生理作用，参与体内许多激素调控及疾病的病理过程。神经肽及其受体的研究已成为当今生理及神经科学研究中最活跃的领域之一。本书共 7 章。第一章和第二章为研究背景和意义，介绍牛卵泡发育过程和 CART 及其相互作用蛋白研究进展，并明确了研究内容、技术路线；其余章节为试验部分，通过牛卵泡转录组的高通量测序和 CART 相互作用蛋白研究，初步筛选 CART 候选受体；对 CART 及其候选受体的同源建模，ZDOCK 分子对接获得 4 个 CART 候选受体，并对 CART 及其候选受体进行荧光定量 PCR 验证。

  本专著较为系统地介绍了作者多年来对 CART 的相关研究成果，思路新颖、资料丰富、内容连贯，反映了神经肽 CART 研究的新进展，可供畜牧兽医专业师生、医学院校师生及医学科研人员参考和查阅之用。

前 言
PREFACE

探索和揭示神经肽的奥妙，是当代自然科学面临的重大挑战之一。自20世纪60年代后期神经肽概念的提出，在神经科学中导致了一场革命性的变化。进入20世纪70年代，由于成功地提纯和鉴定了某些神经肽，特别是相关实验技术如放射免疫测定技术的建立，使神经肽研究得以迅速发展。神经肽是神经元间重要的化学信使，它可发挥神经激素、递质和调质的作用；同时，神经肽在脑部和外部组织广泛分布，反映了神经肽生物学功能的多样性和复杂性，引起了各领域科研工作者的研究兴趣。

可卡因-苯丙胺调节转录肽（CART）是20世纪80年代发现的一种在动物体内广泛分布的下丘脑神经肽，参与人和动物多方面的生理功能：①调节动物食欲和进食量；②调节多种激素的释放；③参与疼痛感觉传导和疼痛反应产生；④在药物成瘾方面发挥作用。2004年，首次在牛卵巢上发现CART mRNA表达，并证明CART对牛卵泡发育产生负调控作用。

《牛卵泡可卡因-苯丙胺调节转录肽（CART）受体的筛选》一书，在明确了CART与动物卵泡发育的关系及CART受体特征的基础上，对CART受体的筛选展开了研究，其中，也反映了本课题组8年多来从事神经肽CART研究工作的情况。本书共7章，具体内容如下。

第一章为牛卵泡发育及可卡因-苯丙胺调节转录肽研究进展，介绍牛卵泡发育过程和CART及其相互作用蛋白以及相关实验技术的研究进展；第二章为研究意义、内容和技术路线及创新点；第三章为牛卵泡转录组的高通量测序，通过结果初步筛选CART候选受体；第四章为CART相互作用蛋白的研究,通过免疫共沉淀及质谱分析技术筛选 CART 候选受体;第五章为 CART 及其候选受体的同源建模,通过 SWISS – MODEL 和 MODELLER 同源建模技术获得 CART 及 38 个候选 G 蛋白偶联受体的分子坐标;第六章为 ZDOCK 分子对接,通过 CART 与候选受体的分子对接,筛选出 4 个 CART 候选受体;第七章为荧光定量 PCR 验证试验。本专著较为系统地介绍了作者多年来对 CART 的相关研究的进展与成果,思路新颖、资料丰富、内容连贯,反映了神经肽 CART 研究的新进展,可供畜牧兽医专业师生、医学院校师生及医学科研人员参考和查阅之用。

在本书编写过程中,得到中国农业科学技术出版社的大力支持;导师吕丽华教授审校了著作的全部内容;美国密歇根州立大学 Gerogre W Smith 教授和西弗吉尼亚大学姚舰波副教授、中国科学院电工研究所潘卫东教授、中国环境科学研究院关潇博士、山西农业大学动物科技学院刘文忠教授、河北农业大学动物科技学院张英杰教授、山东农业大学动物科技学院王建民教授、Oracle 公司软件工程师付晓峰老师在研究过程中给予了悉心的指导;课题组师弟师妹在资料收集、文稿打印等方面做了大量工作。在此,对关心、支持本书出版的同志表示衷心感谢。

由于作者学识水平所限,书中遗漏及错误之处在所难免,敬请广大同行和读者批评指正。

李鹏飞

2015 年 9 月

目 录
CONTENTS

第一章 牛卵泡发育及可卡因-苯丙胺调节转录肽研究进展 …… (1)
 第一节 牛卵泡发育及其研究方法进展 …………………… (1)
 第二节 CART 及其相互作用蛋白研究进展 ……………… (9)

第二章 研究意义、内容和技术路线 ……………………… (22)
 第一节 研究意义 …………………………………………… (22)
 第二节 研究内容 …………………………………………… (23)
 第三节 技术路线 …………………………………………… (24)

第三章 牛卵巢卵泡高通量测序与 GO 分析 …………… (26)
 第一节 概述 ………………………………………………… (26)
 第二节 材料与方法 ………………………………………… (27)
 第三节 结果与分析 ………………………………………… (30)
 第四节 讨论 ………………………………………………… (34)
 小结 ………………………………………………………… (35)

第四章 牛卵泡颗粒细胞 CART 蛋白质组学研究 …… (36)
 第一节 牛卵泡颗粒细胞总蛋白的提取 …………………… (36)
 第二节 免疫共沉淀及质谱分析 CART 相互作用蛋白 …… (39)
 第三节 生物性息学分析 …………………………………… (48)
 小结 ………………………………………………………… (55)

1

第五章　G蛋白偶联受体的筛选及同源建模 ……………… (57)
　第一节　G蛋白偶联受体的预测 ……………………………… (57)
　第二节　CART及G蛋白偶联受体的同源建模 ……………… (63)
　小结 …………………………………………………………… (78)
第六章　CART与筛选蛋白的分子对接研究 ……………… (79)
　小结 …………………………………………………………… (91)
**第七章　CART及其候选受体基因在牛优势卵泡和从属卵泡
　　　　中的表达** ………………………………………………… (92)
　第一节　牛发情周期第一卵泡波优势卵泡和从属卵泡的鉴定 … (92)
　第二节　CART及其候选受体在牛优势卵泡和从属卵泡的
　　　　　表达量分析 ……………………………………………… (100)
　小结 …………………………………………………………… (108)
研究展望 ……………………………………………………… (109)
附：本著作来源于以下已发表或待发表论文研究成果 ……… (111)
参考文献 ……………………………………………………… (115)

第一章

牛卵泡发育及可卡因-苯丙胺调节转录肽研究进展

第一节 牛卵泡发育及其研究方法进展

母畜出生前，卵巢上有许多原始卵泡，但只有少数卵泡能够发育成熟和排卵，绝大多数卵泡发生闭锁和退化。因此，卵泡的绝对数随着年龄的增长而减少。如初生母牛犊有 75 000 个卵泡，10～14 岁时 25 000 个，到 20 岁时只有 3 000 个。而且在一个发情周期内，单胎动物（牛、马或部分羊等）卵巢内一般情况下只有一个成熟卵泡释放卵子并受精，卵巢上大量参与发育的卵泡在其发生过程中闭锁而不能排卵。

牛发情周期内排卵卵泡的数量和质量与其繁殖性能直接相关。卵泡发育是一个连续而复杂的过程。按发育阶段不同，分为原始卵泡、初级卵泡、次级卵泡、三级卵泡和成熟卵泡。雌性哺乳动物卵母细胞生长的同时，周围的体细胞开始增生，逐渐由扁平状变为柱状，由单层变为复层，并分化成颗粒细胞和内膜细胞，这些体细胞的分化增生构成早期卵泡增大的主要原因。卵细胞本身在内分泌激素的调控下有序的生长、发育，相关基因在不同时空准确无误地表达。2007 年 George W Smith 课题组通过试验研究证实了可卡因-苯丙胺调节转录肽（Cocaine - and

牛卵泡可卡因－苯丙胺调节转录肽（CART）受体的筛选

amphetamine – regulated transcript peptide）对牛卵泡发育有着显著的负调控作用，能够引起牛卵泡闭锁，因而是导致牛繁殖力下降的一个重要因素。本章就牛卵泡发育的研究概况及CART与卵泡发育关系的研究进展以及相关的实验技术进行综述。

1960年，"双卵泡波"提出，此后，"优势"卵泡概念、优势卵泡激素调控理论相继建立，并随着放射免疫测定技术、超声波技术、体外细胞培养技术等的成熟和广泛应用，已经明确在牛发情期内优势卵泡生长和闭锁过程通常经历2个或3个"卵泡发育波"；并发现在牛卵泡液中存在许多卵泡内生长因子，通过体外实验研究表明这些生长因子可通过内分泌、自分泌和旁分泌作用，修饰促性腺激素刺激卵泡生长和分化；通过对颗粒细胞凋亡和体外实验研究发现，颗粒细胞凋亡是引起卵泡闭锁的主要原因。然而，卵泡生长和闭锁机理仍不明确。可以确定，随着新技术、新方法的改进和应用，对牛卵泡生长和闭锁机理的研究将会极大改善牛繁殖性能。

对牛排卵和卵泡发育相关的研究中，人工调控发情周期的持续时间、排卵时间和排卵数量一直是研究热点。随着研究技术和手段的发展和应用，卵巢有腔卵泡生长的特定模式相继建立。近年来，许多研究人员致力于牛优势卵泡生长和闭锁规律研究，尤其是促性腺激素与卵泡内生长因子相互作用结果、颗粒细胞凋亡与卵泡闭锁的关系。

一、牛发情周期内优势卵泡发育模式研究进展

（一）"双卵泡波"和"三卵泡波"假设的提出

Rajakoski（1960）报道：在一个发情周期内，直径≥5mm的卵泡在生长过程中出现2个"发育波"，第一个发育波出现在发情周期第3～4d，第二个出现在发情周期第12～14d，二者导致排卵前卵泡形成；第一个卵泡发育波中大卵泡从第12d到第12～17d逐渐衰退闭锁，第二个发育波中大卵泡最后成熟、排卵。因此，Rajakoski是第一个提出在

牛发情周期内卵泡生长和闭锁的"双波"概念。

Hodgen等（1977）提出"优势"的概念解释了灵长类动物月经周期内，当有排卵卵泡或黄体存在时卵泡生长衰退的原因。然而，牛发情周期内"优势卵泡"的存在直到1987年才获得形态学、组织学、激素和生物化学研究的支持，结果表明：在发情周期的3个不同阶段，雌激素活跃卵泡最终发育为最大卵泡；根据这些研究结果，Ireland和Roche得出结论：在牛发情周期内，优势卵泡出现3个阶段，优势卵泡发育每个阶段都经历选择、优势、闭锁或排卵，这与灵长类动物月经周期优势排卵卵泡的描述相同。

（二）牛有腔卵泡分类方法的建立

Ireland等通过对卵泡液中雌激素与孕酮或雄激素的比值首次对牛有腔卵泡（直径≥6mm）进行分类，分为雌激素活跃卵泡和雌激素不活跃卵泡。雌激素活跃卵泡（雌激素浓度＞孕酮或雄激素浓度）具有健康生长卵泡的生物学特性，雌激素不活跃卵泡有闭锁卵泡或注定向闭锁卵泡发育的特性。

Ireland等研究结果还表明：在发情周期的卵泡期和黄体期，雌激素活跃卵泡的生长与颗粒细胞（CGs）增殖、卵泡内雌激素浓度增加、CGs和膜细胞（TGs）中促黄体素（LH）结合位点数量有关。相反，雌激素不活跃卵泡CGs中促卵泡素（FSH）结合位点数量减少。在卵泡阶段，排卵期前卵泡促性腺激素激增或黄体期9~11d的最大卵泡就是雌激素不活跃卵泡。与雌激素活跃卵泡相比，雌激素不活跃卵泡其卵泡内雌激素浓度降低，CGs和TGs中促性腺激素受体位点数量减少。总之，雌激素活跃卵泡将FSH应答能力转向LH应答能力作为发育的基础，CGs和TGs中促性腺激素受体数量减少与雌激素活跃卵泡产生雌激素能力丧失有关。

（三）促性腺激素与卵泡发育的关系

Dobson（1977）在牛排卵期前3d每隔2h通过RIA测定血清促性

腺激素浓度，发现在发情周期内牛血清 FSH 浓度有 2 个瞬时峰值，显著的 FSH 瞬时峰值与排卵期前 LH 峰值相伴随；在促性腺激素峰值出现 12~24h 后，当血清中 LH 浓度保持较低浓度或稳态时，另一个 FSH 波峰出现。Adams 等（1992）对显著的 FSH 波峰与卵泡波的关系加以阐述，在牛发周情期内有两个卵泡波，随着每个卵泡波出现，血清 FSH 浓度显著提高 50%~75%；通过对牛 FSH 浓度和卵泡波出现时间的研究，发现 FSH 波峰在排卵第 1d 和发情周期第 10d 出现；同时，还报道牛有 3 个卵泡波，在每个卵泡波前 1d 出现 FSH 波峰，即在发情周期第 1d、9d 和 15d。这些研究结果表明：在发情周期第一卵泡波和第二卵泡波之前，FSH 浓度存在显著的瞬时双倍增加，如图 1-1 所示。

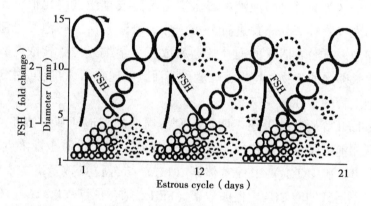

图 1-1　牛发情周期内每个卵泡波与 FSH 瞬时峰值关系的模型

（引自 James J Ireland et al., 2000）

Figure 1-1　A model depicting the relationship of transient peaks of FSH with each wave of follicular development during the estrous cycle of cattle

为了探究 FSH 在牛卵泡波形成中所起作用，Miller 等（1979）证明在牛卵泡期注射牛卵泡液，可延迟发情期到来。Good 等（1995）也证实牛卵泡液是一种来源丰富的抑制剂，对切除卵巢和未切除卵巢的牛，抑制 FSH 分泌而不改变 LH 分泌。总之以上研究结果充分证明在牛发情周期内 FSH 瞬时升高能刺激卵泡波内卵泡生长。

第一章
牛卵泡发育及可卡因－苯丙胺调节转录肽研究进展

FSH 浓度瞬时升高引发卵泡波，同时很多研究也表明，在一个卵泡波内，FSH 浓度降低能够刺激优势卵泡发育以及选择优势卵泡。Adams 等（1993）证明在一个卵泡波内，当 FSH 浓度从波峰自然降低时，给牛注射 FSH 可延迟优势出现，后来得到 Mihm 等研究证实。这些研究结果表明：FSH 浓度从波峰下降过程中对选择过程终止和刺激优势出现起重要作用。

Rohe 等（1980）证明在牛发情期内 LH 脉冲分泌模式。牛颈静脉插管，在发情期第 3d、10d 或 11d、18d 或 19d，24h 内每隔 10min 采血，结果显示，在黄体期，LH 分泌模式由黄体期前期低振幅高频率脉冲改变为中期高振幅低频率；在发情期，LH 脉冲分泌模式转变为高频率（类似于黄体期前期）很高振幅的脉冲。Rohe 等的研究结果随后得到其他课题组研究证实。Cupp 等（1995）对这些研究内容加以充实：发情周期黄体期内，在孕酮浓度较高的不同时间 LH 分泌频率和振幅显著发生改变。然而，上述研究都没有对发情周期内优势卵泡生长和闭锁的动力学与 LH 脉冲分泌模式改变之间的关系进行研究。

Evans 等（1997）首次对牛发情周期内 LH 分泌脉冲动力学改变与卵泡波相互关系进行研究。他证明在第一个卵泡波优势卵泡生长期间，LH 分泌频率降低但振幅增加；与此相反，在第一个卵泡波优势卵泡丧失优势、第二个卵泡波生长卵泡出现后，LH 分泌频率保持不变，但振幅降低至发情周期第 3～4d 的水平。尽管 Cupp 等在研究过程中没有证明 LH 分泌频率和振幅的改变与卵泡波之间的关系，但研究结果与 Evans 等的结论相似，即发情周期卵泡波中 LH 脉冲分泌模式是不同的。

发情周期内 LH 脉冲分泌模式的改变调节优势卵泡生长和闭锁仍有待于研究。然而，可以肯定的是 LH 脉冲分泌模式的改变对优势卵泡生长和闭锁的调节有重要作用。优势卵泡的发育依赖于 FSH 和 LH 浓度的转变。因此，LH 脉冲分泌频率的降低，在发情周期一个卵泡波的后期，不仅与每一个卵泡波内优势卵泡优势的丧失有关,而且可能是引起优势丧失的原因。

(四)卵泡内因子与颗粒细胞凋亡对卵泡发育的影响

卵泡内因子尤其是生长因子如:抑制素、激活素、胰岛素样生长因子Ⅰ和Ⅱ(IGF-Ⅰ,IGF-Ⅱ)及其结合蛋白,对卵泡生长分化和功能调节有重要作用。除内分泌功能外,抑制素、激活素、IGF、卵泡抑素及IGF结合蛋白自分泌和旁分泌活性具有刺激或抑制颗粒细胞和膜细胞的功能。此外,各因子间在功能上存在颉颃作用,如抑制素与激活素、激活素与卵泡抑素、IGF与IGF结合蛋白。由于这些因子及结合蛋白存在于牛卵泡液,各生长因子及其结合蛋白在卵泡内含量比例可能会发生变化,进而影响卵泡生长、分化及功能,但体外试验验证较为困难。因此,准确解释促性腺激素与卵泡内生长因子相互作用调节卵泡波内优势卵泡生长和闭锁的机制仍需进一步研究。

颗粒细胞的增殖分泌功能以及与内分泌激素相互影响,对维持卵泡正常生长、发育和成熟起着非常重要的作用。Pederson等(2003)研究发现,闭锁卵泡总伴随着细胞严重凋亡,随着卵泡中颗粒细胞逐渐发生凋亡,卵泡随之发生闭锁。Bilodeau等(2002)也证实严重闭锁卵泡中卵母细胞发育能力很低。Wongsrikeao等(2005)认为颗粒细胞的包裹对卵母细胞生发泡破裂(GVBD)有关键作用,与颗粒细胞共培养有助于卵母细胞完全成熟,并有助于防止卵母细胞DNA片断化。目前对卵泡液中游离颗粒细胞研究较少,对颗粒细胞密度与卵泡闭锁及卵母细胞发育能力之间的关系尚无报道。Yang和Rajamahendran(2000)研究发现,牛卵泡液中游离颗粒细胞凋亡比膜细胞更严重,Zeuner等(2003)的试验也观察到类似的结果。Luciano等(2000)认为,颗粒细胞凋亡后细胞连接遭到破坏,凋亡细胞进入卵泡液中,从而使卵泡液中凋亡的颗粒细胞密度增加。因此,卵泡液中游离颗粒细胞的密度有可能反映了卵泡闭锁程度及卵母细胞的发育能力,从而更加简便、快捷地挑选出高质量卵母细胞。

Yao 等（2004）首次在牛卵巢上发现 CART mRNA 表达，主要在卵泡颗粒细胞表达。Yasuhiro 等（2004）证明 CART 在牛卵泡颗粒细胞上有表达，Sen 等（2007）和吕丽华等（2009）通过研究进一步证明 CART 在一定 FSH 浓度下，对颗粒细胞雌激素的分泌有抑制作用，并对卵泡发育有着显著的负调控作用，能够引起牛卵泡闭锁，是导致牛繁殖力下降的一个重要因素。总之，这些研究表明，颗粒细胞凋亡或受到抑制，最终会导致卵泡闭锁。

（五）利用牛发情周期黄体的特征识别卵泡波

McNutt（1924）和 Hammond（1927）报道在牛发情周期内黄体的形态上有明显的变化，尽管黄体的外观通常用来判断发情周期的阶段，尤其是牛，但在很多物种上判断的准确性仍未报道。Ireland 等（1980）通过对黄体表观总的变化特征的研究，提出了准确判断母牛发情阶段的方法。

在 Ireland 等的研究中，选择未知年龄海福特小母牛 180 头判断发情周期的阶段，并将这些特征分成 4 个阶段，描述见表 1-1 所示。通过与已知的发情周期天数对比，确定了准确判断发情周期 4 个阶段的方法。

表 1-1 牛黄体发育的形态学特征

Table 1-1 Morphological features of bovine corpus luteum development

分类	发情周期阶段（d）			
	Ⅰ (1~4)	Ⅱ (5~10)	Ⅲ (11~17)	Ⅳ (18~20)
黄体的外观（外部）	红，近日排卵；裂点未被上皮细胞覆盖	裂点被上皮细胞覆盖，黄体顶点呈红或褐色	黄褐色或橙色	亮蓝色→白色
内部	红色，偶尔充满血液，细胞松弛，有规则	仅顶点呈红或褐色，其余黄体呈橙色	橙色	橙色→黄色
直径	0.5~1.5 cm	1.6~2 cm	1.6~2 cm	<1 cm

(续表)

分类	发情周期阶段 (d)			
	Ⅰ (1~4)	Ⅱ (5~10)	Ⅲ (11~17)	Ⅳ (18~20)
黄体表面脉管系统	不可见	通常外围固有	与阶段Ⅱ相同，但在后期黄体的顶点将被覆盖	不可见
卵泡直径>10mm	不可见	出现	可能出现或不出现	出现

（引自 James J Ireland et al., 1980）

第Ⅰ至第Ⅳ阶段的时间长短相对于发情周期是未知的。因此，评价这4个阶段相对于发情周期的时间长短，母牛在每一阶段总数的百分数是随母牛发情周期的平均长度（20d）增加的。优势的非排卵卵泡存在的持续时间是从优势卵泡出现至消失的间隔时间，对第一、第二卵泡波优势卵泡的持续时间是7.4~13.1d、11.4~17d。每个卵泡波中直径≥5mm的非优势卵泡的平均数量是5.3个、3.9个和4.7个，持续时间大约6d，最大直径通常≤8mm。对于只有两个卵泡波的牛，第一和第二卵泡波出现于发情周期的第2~4d和第10~11d，在第6d和第19d分别达最大直径约13mm。排卵卵泡在排卵前10d左右能被检测到。

因此，通过对牛发情周期内黄体外观形态变化，推测出该黄体所处发情周期天数，结合不同卵泡波所处的发情周期天数，可以推测出卵泡波所处阶段。

二、卵泡发育研究方法的研究进展

Dufour等（1972）和Matton等（1981）根据牛卵巢表面有腔卵泡大小不同提出的试验方案是选择多组处于发情周期不同天数的牛，通过外科手术，在每对卵巢最大卵泡或每个卵巢上两个最大卵泡的外周间质组织中注入墨汁点，标记3~5d后，通过卵巢切除术检查卵泡大小和发育情况。

以往研究牛发情周期内卵泡生长和闭锁的动力学主要通过屠宰牛或

第一章
牛卵泡发育及可卡因-苯丙胺调节转录肽研究进展

切除卵巢后对牛卵泡形态学进行研究，尤其是对卵泡发育过程进行研究，至少需要两次外科手术。1984年，有两个课题组报道通过超声波扫描仪对奶牛和小母牛的卵巢结构进行检查。然而，Pierson等（1984）是首次通过超声波扫描术来检测牛整个发情周期卵泡的直径。超声波扫描术所获得的卵泡直径通常是指卵泡无反射波空腔的宽度，里面充满卵泡液，因此，通过超声波扫描术获得的卵泡直径要比同一卵泡通过外科手术获得的直径小2~3mm。

1988年，有3个课题组公布了4种较相近的超声波扫描研究方案，即在牛整个发情周期每天用超声波分析监测个体卵泡发育情况，与以往研究不同之处在于：在每次超声波检测中，卵巢前极和后极、每个卵巢的卵巢大弯和门区、黄体以及直径≥5mm卵泡的具体位置均被确认并标注在示意图上。这样，通过每天超声波分析结果和卵巢示意图，就明确了一个发情周期内卵泡生长和闭锁的模式。

Sirois和Fortune（1990）报道，将孕酮释放装置置于牛阴道中段，可显著提高卵泡寿命和优势以及自发黄体消融后优势卵泡雌激素的产生。与此相反，将2个孕酮释放装置置于阴道内模拟黄体期孕酮水平，结果不影响卵泡寿命以及优势卵泡生长和闭锁。Savio等（1993）通过阴道内孕酮释放装置或埋植诺斯孕甾酮人工控制血清孕酮浓度，在发情周期每天或双数天用超声波技术监测优势卵泡发育。

近年来，研究者通过体外细胞培养技术，对CART与卵巢卵泡颗粒细胞激素分泌和调控进行了研究，发现卵泡颗粒细胞雌激素的分泌需在FSH诱导下产生，并对卵泡发育有负调控作用，能够引起牛卵泡闭锁。

第二节 CART及其相互作用蛋白研究进展

一、CART的研究进展

CART是近年来发现的一种在动物体内广泛分布的内源性神经肽，参与人和动物多方面的生理功能：①CART调节动物食欲和进食量，与

牛卵泡可卡因-苯丙胺调节转录肽（CART）受体的筛选

肥胖的发生有密切关系。②CART可以调节下丘脑—垂体—甲状腺轴、下丘脑—垂体—肾上腺轴、下丘脑—垂体—性腺轴等从而调节多种激素释放。此外，CART还能通过与其他激素或其受体共表达来调节激素释放和作用。③近年来对其功能进一步研究发现CART可能在调节疼痛过程中起重要作用，如参与疼痛感觉传导和疼痛反应产生，与其他疼痛相关神经递质（调质）的相互调节作用等。④CART在药物成瘾方面发挥作用。⑤已有学者对牛的研究中已经证实CART对卵泡健康状况有负调节作用，并且CART对牛颗粒细胞产生的雌激素基本含量也有负调控作用。目前对CART功能研究主要集中在对牛繁殖性能的影响。

Douglass等（1995）在大鼠腹侧纹状体内发现CART mRNA之后，有研究者运用定向标记PCR扣除杂交法进一步证实，CART mRNA是下丘脑中含量最丰富的mRNA之一。在随后的研究中还发现CART及CART mRNA也存在于其他脑区及组织中。到目前为止，大鼠、小鼠、金鱼以及人类的CART基因已得到分离并测序，它们均由3个外显子和2个内含子组成，C端由3对二硫键保持其三维空间构型，这是CART活性所必需的，N端构象则缺乏规律性。在研究CART序列及其加工处理过程中发现CART有两种长度的前体，分别是长型（129个氨基酸）和短型（116个氨基酸）。大鼠和小鼠体内都有长型CART、短型CART两种，其中短型CART由89个氨基酸残基组成，长型CART是在短型CART的第39位氨基酸残基的C端插入另外13个氨基酸残基构成，总长度为102个氨基酸，其前体蛋白均含有27个氨基酸残基组成的疏水性前导序列。

大量研究表明，CART能通过促肾上腺皮质激素释放激素（Corticotropin releasing hormone，CRH）途径参与调节下丘脑-垂体-肾上腺轴（Hypothalamic pituitary adrenal，HPA）摄食调控的功能。通过动物活体外大量研究结果发现，CART能对细胞培养下丘脑产生刺激并使其释放CRH；通过对动物活体研究也发现，脑室内灌注重组体CART（42~89）活性片段后，下丘脑室旁核（Paraventricular hypothalamic nucleus，

PVN）内肾上腺皮质激素释放激素神经元早期快反应基因 $c-fos$ 的表达量增加，同时在荧光显微技术观察发现，PVN 内 $c-fos$ 表达增加，肾上腺皮质激素释放激素神经元与 CART 免疫反应神经元临近，在同一时期还伴随体内催产素水平的暂时性升高以及皮质酮、血糖水平的显著升高。研究发现，CART 可以通过调节下丘脑 - 垂体 - 甲状腺轴来增加产热。在体外培养中，下丘脑中的促甲状腺激素释放激素（Thyrotrop(h)in - releasing hormone，TRH）受到 CART 的调节作用而表现为 TRH 水平增加，但也有相关的研究表明在脑室内给予 CART 并没有引起外周血中促甲状腺激素（Thyrotropic - stimulating hormone，TSH）及 TRH 含量变化。

CART 可调节垂体激素的释放。在垂体前叶，CART 位于促性腺细胞内，可以抑制催乳素（Prolactin，PRL）释放。而 CART 与 TRH 只在室旁核促垂体神经元内共存，对于培养的垂体前叶细胞，单独使用 CART，TSH 和 PRL 释放无改变，但 CART 可以剂量依赖性地减少 TRH 诱导的 PRL 释放，而对 TRH 诱导 TSH 释放无影响。提示 CART 对 TRH 诱导 PRL 有重要作用。含有 CART 的神经元末端能与促垂体区的 CRH 和 TRH 神经元发生突触联系；在体外培养情况下，CART 能促使下丘脑中的 CRH、TRH 和神经肽 Y（Neuropeptide Y，NPY）水平增加。

动物采食量与体内血液葡萄糖浓度之差呈负相关，而器官或细胞的吸收率是体内血液葡萄糖浓度差异的标识，细胞吸收利用葡萄糖的能力会受到体内胰岛素水平的影响。同时还有研究证实，在对中枢长期灌注 CART 后，动物表现为采食量和体重下降，此外还伴随血糖水平上升和胰岛素水平下降。总之，研究表明，CART 很可能通过对体内胰岛素水平的调节进一步影响血糖浓度之差来调节动物采食量。

瘦素和 NPY 是两种食欲调节因子，而食欲调节因子又是中枢食欲控制系统中重要的组成元素。弓状核可以合成 CART、NPY、阿黑皮素原（Proopiomelanocortin，POMC）和刺鼠相关蛋白（AGRP）。瘦素则主要通过作用于弓状核产生相应的进食或抑食控制，由此来调节体内能量

牛卵泡可卡因－苯丙胺调节转录肽（CART）受体的筛选

代谢状况，从而进一步调节脂肪储存量，以保持体内能量相对平衡。NPY 是一种重要的中枢摄食刺激因子，它能够调节体内能量摄入和支出。由于 NPY 这种重要的神经调节作用，似乎所有与摄食抑制有关的神经肽都会与 NPY 相互作用来共同影响动物采食。在体循环中，葡萄糖、胰岛素和瘦素水平是体内能量平衡的重要信号，但向中枢传递体内能量储存情况的主要信号还是瘦素。

穹窿核和室旁核与中枢自主神经的输出投射均有联系，其中，可以投射到其他与食欲相关的边缘系统结构中，还可以直接将整合信号传送至外周。所传送信号在下丘脑核团中进行整合，低位脑干与下丘脑之间也通过相互作用对进食行为进行调节。传入外周的信号通过调整胃肠排空速率、胃酸分泌量、胆囊收缩素饱腻感及产热等对进食及体内能量状况进行调节。当此类与进食相关的中枢功能异常时，大脑将不能及时且正确地对来自各个方面调节进食的激素做出相应的调节反应，就会引起进食行为与身体能量需求状况的不平衡，最后引起肥胖。由此可见，CART 是体能调节通路中受瘦素调节的重要激素之一，在进食行为和体能平衡调节过程中起着重要作用。

（一）家猪卵巢卵泡 CART 的研究概况

本课题组是国内外首次对家猪 CART 进行研究。目前，主要从 CART mRNA 的表达和 CART 对卵泡颗粒细胞雌激素产生的影响入手开展研究。

研究过程中（2010）选取 5 头雌性大白猪（山西省晋中市太谷县屠宰场）组成试验群。待母猪屠宰后，取出下丘脑、卵巢，进行总 RNA 提取。设计 3 对引物进行 PCR 扩增。结果显示，从下丘脑、卵巢总 RNA 中均获得片段大小为 200 bp、250 bp 与 500 bp 左右的序列，经拼接后获得的 CART 全长 797 bp，其中包含完整的开放阅读框（Open reading frame，ORF），Blast 分析为 CART 的完整 CDS 区，共编码 116 个氨基酸。经同源序列相似性分析，结果显示：在核苷酸水平，家猪与野猪、牛、小鼠、犬的 CART 之间有较高的相似性（88%~99%）；在氨基酸水平，家猪与其他物种 CART 氨基酸序列有很高的相似性，其中，与野猪、牛的相似性达100%；与小鼠、人、黑猩猩、猕猴、犬的 CART 相似性分别为97%、94%、95%、96%和92%。

通过生物信息学分析，家猪 CART 编码蛋白质的 3D 结构呈不规则扭曲状，具有 3 对共价键参与了三维空间构型的维持。对 CART 蛋白质序列分析发现，在 C 端附近存在编码 6 个半胱胺酸的序列，经同源性比对发现，每个物种 CART 均存在编码 6 个保守的形成二硫键的半胱胺酸序列。说明该区域形成 CART 的序列模体，同时该二硫键模体的存在对 CART 3D 框架结构的形成具有重要意义。通过结构域分析，家猪 CART 蛋白质在第 40~116 位氨基酸间存在一个 CART 超家族（CART superfamily）。

家猪卵巢卵泡在发育过程中，小卵泡为卵泡发育的初始阶段，中卵泡为卵泡走向闭锁或是走向优势卵泡的分化阶段，大卵泡则会走向成熟排卵。在家猪卵巢卵泡 CART mRNA 表达量的研究中，将猪卵泡按直径大小划分为大卵泡（d>5mm）、中卵泡（3mm<d<5mm）、小卵泡（d<3mm）3 个等级，利用 qRT-PCR 技术来检测 CART mRNA 在猪卵巢中大、中、小卵泡内的表达是否有差异。结果显示 CART mRNA 在中等卵泡内表达量显著高于大、小卵泡（$P<0.05$），表明猪卵巢中中卵泡是卵泡走向优势卵泡或闭锁卵泡的转折点，CART mRNA 在中卵泡内表达量最高，说明在猪卵泡发育的关键阶段，CART 起到了调控作用。

同时，采集 6 个大、中、小卵泡，采用双抗体夹心 ABC-ELISA 法，选用雌二醇放射免疫测定试剂盒测定卵泡内雌激素（Estrogen，E_2）和孕酮（Progesterone，P）含量。测定结果经显著性检验，大、中、小卵泡内 E_2 和 P 浓度差异均不显著（$P>0.05$）。这表明，在猪卵巢卵泡中 CART 对卵泡颗粒细胞 E_2 和 P 的分泌没有直接影响。为了深入了解 CART 对猪卵泡颗粒细胞生长的影响，我们进行了体外培养研究。采用 Long-term 培养方法,CART 选择 4 个浓度梯度(0μmol/L、0.01μmol/L、0.1μmol/L、1μmol/L),FSH 选择 4 个浓度梯度(0ng/ml、5ng/ml、25ng/ml、50ng/ml),37℃,5% CO_2,饱和湿度条件下,在 MEM-α 培养液里培养 7d,用 DPBS(Dulbecco PBS)胰蛋白酶消化并细胞计数。分别在培养 48h、96h、144h 和 168h 后,在显微镜下观察各孔细胞生长情况并采集图像;培养 168h 后,进行 E_2 测定。结果经显著性检验表明:当培养体系 FSH 浓度分别为 25ng/ml 和 50ng/ml 时,随着 CART 浓度的升高(0.01μmol/L、0.1μmol/L、1μmol/L),培养液 E_2 浓度有下降趋势,但各组间差异不显著($P>0.05$)。但是,当 FSH25ng/ml,CART 0.1μmol/L 和 1μmol/L 时,与对照组相比,颗粒细胞 E_2 的产生明显受到抑制($P<0.05$)。但是,统计分析 FSH 和 CART 互作效应并不显著($P>0.05$)。

(二)绵羊卵巢卵泡 CART 的研究概况

在研究 CART mRNA 在绵羊卵巢卵泡颗粒细胞表达过程中,以绵羊下丘脑组织中的 CART 基因作为阳性对照,空白组作为阴性对照。研究结果表明,CART 基因在绵羊卵巢卵泡颗粒细胞中有表达,该片段长度为 255 bp,绵羊卵巢卵泡颗粒细胞和下丘脑的 CART 基因 cDNA 序列相似性为 100%,将测序结果与 GenBank 中收录的牛、马、人类、黑猩猩、野猪和猕猴等动物的 CART 基因 cDNA 序列进行相似性分析,该片段与牛(NM 001007820)、野猪(AF338229)、沙漠刺毛鼠(NM EU477391)、黑猩猩(XM001155069)、人类(NM 004291)、马(XM 001504033)、猕猴(XM 001096906)的 CART 基因序列相似性分别为 98%、94%、92%、90%、90%、91% 和 89%。

我们进一步通过体外培养绵羊卵泡颗粒细胞来研究 CART 对卵泡发育的影响。采用 Long-term 培养方法,CART 选择 4 个浓度梯度(0μmol/L、0.01μmol/L、0.1μmol/L、1μmol/L),FSH 选择 4 个浓度梯度(0ng/ml、5ng/ml、25ng/ml、50ng/ml),体外培养 168h 后,对颗粒细胞生长情况的图像采集显示,当 FSH 浓度为 0ng/ml 时,随着 CART 浓度的逐渐增大,卵巢卵泡颗粒细胞增殖在各组没有差异,而且细胞稀疏松散,没有成簇细胞群。表明绵羊卵巢卵泡颗粒细胞的增殖需在 FSH 诱导下进行,这时 CART 对其影响不大。当 CART 浓度为 0μmol/L 时,随着 FSH 浓度逐渐加大,卵巢卵泡颗粒细胞密度也逐渐增大。表明 FSH 对体外培养的绵羊卵巢卵泡颗粒细胞的增殖有促进作用。当 CART 和 FSH 同时存在时,对绵羊卵巢卵泡颗粒细胞体外培养 168h 后,颗粒细胞生长情况的图像显示,FSH 浓度分别为 0ng/ml、5ng/ml、50ng/ml 时,随着 CART 浓度的增大,卵巢卵泡颗粒细胞密度逐渐减小。但当 FSH 浓度为 25ng/ml、所加 CART 浓度为 0.01μmol/L 时,细胞明显松散,没有成簇细胞团出现。

体外培养 168h 后,E_2 测定结果经 SPSS 单变量线性模型分析显示:FSH 主效应为极显著($P<0.01$)、CART 主效应极显著($P<0.01$)、互作效应极显著($P<0.01$),说明 FSH 和 CART 之间有交互作用。当 FSH 浓度为 5ng/ml,培养液添加不同浓度 CART(0.01μmol/L、0.1μmol/L、1μmol/L),颗粒细胞 E_2 的产生受到抑制,其中,CART 浓度为 0.1μmol/L 时,培养液 E_2 浓度最低。当培养液 FSH 浓度分别为 25ng/ml、CART 浓度为 0μmol/L、0.01μmol/L 时,培养液 E_2 浓度较高,但是随着 CART 浓度继续增加为 0.1μmol/L 和 1μmol/L 时,其对颗粒细胞 E_2 的产生明显抑制作用($P<0.05$),当培养液 FSH 浓度分别为 50ng/ml 时,CART 浓度在 0μmol/L、0.01μmol/L 时,对培养液 E_2 产生的抑制作用不明显;随着 CART 浓度继续增加为 0.1μmol/L 时,培养液 E_2 浓度显著降($P<0.01$)且达到最低。

总之,研究表明,绵羊卵巢卵泡颗粒细胞 E_2 的产生是在 FSH 诱导下进行。在试验设计的培养体系中,FSH 浓度为 5ng/ml 时,颗粒细胞产生的 E_2 浓度最大;结果还表明,CART 对绵羊卵巢卵泡颗粒细胞 E_2 的产生有抑制作用,特别当培养体系中加入 CART 浓度为 0.1μmol/L 时,对 FSH 诱导的绵羊卵巢卵泡颗粒细胞 E_2 产生的抑制作用显著。

（三）蛋鸡卵巢卵泡 CART 的研究概况

2012 年，于雪静对蛋鸡卵巢卵泡 CART mRNA 表达进行了研究。试验中选取蛋鸡小白卵泡（Small white follicle, SWF, d < 2mm）提取总 RNA，设计 CART 特异性引物 PCR 扩增，产物经过连接转化、蓝白斑筛选后，测序获得 230 bp 片段。经 Blast 比对发现，该片段与牛同源序列有 3 个碱基差异，与人同源序列存在 19 个碱基差异，与猪同源序列存在 11 个碱基差异，表明所获得的片段为 CART 部分序列，说明 CART mRNA 在蛋鸡卵泡上有表达。

在此基础上，采集不同蛋鸡 8 个不同直径大小的卵泡。卵泡按直径大小可划分为成熟卵泡和未成熟卵泡。成熟卵泡，即排卵前卵泡，分别表示为 F1、F2、F3、F4……未成熟卵泡即小卵泡，可划分为 SWF（d < 2mm）、大白卵泡（Yellow white follicle, LWF, 3~5mm）、小黄卵泡（Small yellow follicle, SYF, 6~8mm）和大黄卵泡（Large yellow follicle, LYF, 9~12mm）。分离 GCs 和 TCs，分别提取总 RNA 进行 qRT-PCR 检测，结果经 SPSS 软件进行单因素方差分析和显著性检验。结果显示：直径 6~8mm 卵泡颗粒细胞内 CART mRNA 表达量显著高于其他卵泡（$P < 0.05$）。同时，为了更准确地定位表达部位，采用免疫组织化学技术对蛋鸡卵巢卵泡 CART 进行定位研究。采集蛋鸡 SWF 和 LWF，组织固定后，经免疫组化结果显示：CART 在蛋鸡卵巢卵泡颗粒层细胞、膜层细胞和卵母细胞均有表达，且膜层表达量高于颗粒层细胞。

（四）牛卵巢卵泡 CART 的研究概况

Kobayashi Y 等（2004）首次在牛卵巢中发现 CART 有表达。通过设计引物成功获得 728 bp 的 CART 片段，并且通过免疫组化、原位杂交等技术将其定位于有腔卵泡的卵母细胞、颗粒细胞和卵丘细胞中；深入研究发现在第一卵泡波，CART mRNA 和卵泡液中 CART 在 SF 比在

第一章
牛卵泡发育及可卡因-苯丙胺调节转录肽研究进展

DF多，且某些剂量CART处理颗粒细胞时发现，CART能引起E_2的产生受到抑制，但对P的分泌量无影响，其效果依赖于细胞分化所处的阶段，这表明CART可能对卵泡的闭锁有调节作用。

Aritro等（2007）为了解释CART是如何抑制FSH诱导的牛卵巢卵泡颗粒细胞E_2的分泌进而对牛卵泡发育产生负调控作用，对牛卵泡上CART的信号传导通路进行了研究，结果表明：CART高度抑制了FSH诱导的卵泡颗粒细胞cAMP水平的升高，并抑制了E_2的积累以及芳香化酶mRNA的表达。此外，研究结果还表明，Ca^{2+}对FSH诱导的卵泡颗粒细胞cAMP水平的升高以及E_2的积累是必需的，而CART能明显抑制FSH诱导的Ca^{2+}流入。通常选择G蛋白和蛋白激酶抑制剂研究CART的信号通路，CART对FSH信号和E_2生成的抑制效应通过Go/i依赖的信号通路调节，而没有其他的抑制信号对CART活性起作用。总之，研究证明了CART对FSH信号通路的多个环节、E_2生成和卵泡发育及闭锁发挥作用，对深刻理解CART潜在的对内外生殖系统的作用机制具有重要意义。

在对CART信号传导通路研究的基础上，Aritro、吕丽华等（2008）提出了牛卵泡颗粒细胞内CART作用机制的模型。在该模型中，FSH刺激Erk1/2和Akt激活，CART则加速FSH诱导的Erk1/2和Akt信号的终止，CART单独刺激并诱导牛卵泡颗粒细胞Erk1/2激活过程中Go/i-PKC-MEK依赖的通路。是否CART诱导Go/i蛋白直接使PKC磷酸化，然后激活MEK和Erk1/2，或者是否Erk1/2通过PKC依赖的方式被激活，这些机理仍不清楚。然而，CART刺激Erk1/2激活是不依赖于PKA、PI3K、和RTK，双特异性磷酸酶5（Dual specificity phosphatase 5，DUSP5）的表达依赖于Erk1/2。因此，提出的负反馈环中，活化的Erk1/2诱导DUSP5表达，反过来终止Erk1/2激活。CART诱导DUSP5表达，功能上需要CART诱导Erk1/2信号的终止和通过Erk1/2依赖的转录机制的调节。CART提高PP2A表达同样功能上需要Erk1/2和Akt活化的终止，CART调节PP2A表达是通过转录依赖和非依赖的机制调节的。

吕丽华等（2009）通过对比出现偏差后的 DF 和 SF，检测出 SF 内 CART 含量高于 DF 内 CART 含量；对早期的优势卵泡注射 CART 后，发现其内的 E_2 浓度下降，且颗粒细胞中的芳构化酶 *CYP19A1* mRNA 表达量降低，进一步证明了 CART 可引起单胎动物的卵泡闭锁。但到目前为止，对 CART 的作用受体及二者的作用机制还未知，有待作进一步的研究和阐释。

二、CART 相互作用蛋白的研究概况

CART 在 20 年前就曾被提出来研究，生物学功能尚不清楚。经过近十年的研究才初步对 CART 有所了解。CART 作为神经递质与进食、体重、药物滥用、应激以及神经内分泌控制有关。Sen 等（2007）证明，CART 在牛卵泡颗粒细胞上有表达，并通过研究证明 CART 在一定 FSH 浓度下，对体外培养的卵泡颗粒细胞生长有着显著影响。本课题组通过前期研究已经证明，CART mRNA 同样在猪的卵泡颗粒细胞中有表达，初步证明 CART 对颗粒细胞雌激素的分泌产生影响。尽管 CART 的作用及细胞活性已得到明确证实，但是经多年研究与 CART 结合的受体却尚未发现。

（一）CART 相互作用蛋白的研究进展

陈娟（2005）通过建立大鼠脑 cDNA 文库，以 $CART_{41-102}$ 为诱饵，应用细菌双杂交技术筛选与 CART 相互作用蛋白，将得到的阳性结果进行 DNA 测序，并进一步根据生物信息学方法分析阳性结果。获得 6 个可能与 CART 相互作用的已知蛋白，同时获得 22 个可能与 CART 相互作用的未知新蛋白的 DNA 序列，它们与 22 个已知膜蛋白或受体蛋白有短的极相似的片段。但试验最终没有明确找到 CART 的受体。

(二) CART 受体的研究进展

CART 在物种间进化是保守的。大鼠和小鼠的 ProCART 是相同的，由 102 个氨基酸组成，并且被加工成两个主要的活性 CART，即 $CART_{55-102}$ 和 $CART_{62-102}$，经常被用于 CART 的研究。与大鼠和小鼠的 ProCART 相比，人类 ProCART 缺少 27~39 序列，并且在第 55 位存在缬氨酸取代异亮氨酸。CART 是下丘脑最丰富的转录模板之一，而且 CART mRNA 和 CART 免疫反应在整个脑、垂体、肾上腺、胰岛细胞、朗格汉斯细胞和肠道的特殊核子中被发现。然而，到目前还没有 CART 受体被成功克隆。Vicentic 等（2005）报道，$^{125}I-CART_{62-102}$ 和小鼠垂体瘤细胞系 AtT20 能够特异性结合。Keller 等（2006）描述融合蛋白能够与 $CART_{55-102}$ 和绿色荧光蛋白质分离的下丘脑细胞结合。由此表明 CART 可能在下丘脑-垂体-肾上腺轴中发挥作用。

Lakatos 等（2005）研究发现，$CART_{55-102}$ 诱导细胞系 AtT20 细胞外信号调节激酶（Erk-1 和 Erk-2）的激活是通过 G 蛋白偶联受体介导的。因此，Lakatos 等研究利用 AtT20 细胞识别 CART 受体的特异性结合。研究表明，AtT20 细胞内存在一个明显的 CART 受体，这是对 CART 受体特异性结合的第一次证明。成功关键在于选择一块能够显示 $CART_{55-102}$ 信号的组织并且能够保证缓冲液和培养条件的准确。早期研究曾提到大脑中存在受体，这说明 CART 很活跃，能够抑制电压依赖性细胞内 Ca^{2+} 信号并诱导 $c-fos$ 基因的表达。另外，由于在各种测试中 $CART_{55-102}$ 和 $CART_{62-102}$ 的相对效力不同，所以，可能存在多个 CART 受体。在任何情况下，将来试验都应进一步阐述 CART 受体结合的特征，也应更有利于 CART 受体 cDNA 的克隆。由于 CART 在进食、应激和精神兴奋剂药物的作用机能与转化过程中的作用，CART 受体的发现将有利于作为治疗剂，以模仿或阻断内源性 CART 分子的研究。

对于 CART 受体的研究，用放射性标记 CART 的受体结合试验大多不成功，因为碘化反应中易使活性配位体的结合位点分解。Vicentic 等

牛卵泡可卡因－苯丙胺调节转录肽（CART）受体的筛选

（2005，2006）证明，$^{125}I-CART_{61-102}$能够特异性结合到AtT20细胞，这种特异性结合通过一个假定的G-蛋白偶联受体与CART-诱导活化的细胞外信号调节激酶（Erk）途径相联系。Lakatos等（2005）研究Erk磷酸化过程中，嗜铬细胞瘤PC12中$CART_{55-102}$的作用忽略不计。然而，有报道称嗜铬细胞肾上腺髓质中存在密集的CART免疫，可以预期在PC12细胞中也存在CART免疫，因为这些细胞系都来自嗜铬细胞并保持原来的特征。在与细胞膜和完整的游离PC12细胞竞争性结合实验中，经15min达到平衡结合并在随后至少60min保持稳定。在一个低浓度范围内，$CART_{61-102}$、$CART_{55-102}$和I_2-CART_{61-102}以相似的亲和性与细胞膜结合。

在所测细胞膜和非分化及分化PC12细胞所有配体中，二碘化衍生物CART有最高的配体结合亲和性。与非碘化化合物相比，碘化化合物具有较高的受体亲和力。碘化后配体亲和性加强在文献中已有描述，例如对于特定的催产素受体拮抗剂。

Lenka等研究还证明，$^{125}I-CART_{61-102}$与分化的PC12细胞特异性结合，与非分化细胞相比，分化细胞显示更低的非特异性结合（20%：40%）。Vicentic等报道，$^{125}I-CART_{61-102}$在AtT20细胞系中也发生特异性结合。Lenka对此的解释是AtT20细胞系的结合实验在4℃下完成，而PC12细胞的结合实验在37℃下完成。有趣的是，对于分化的细胞，Bmax为11 194±261结合位点/细胞高出非分化细胞Bmax为2 228±529结合位点/细胞的5倍。对非分化和分化PC12细胞中CART结合位点的检测指出CART可能在交感-肾上腺髓质系统和应激反应中起作用。在Lenka等的研究中没有成功地检测到任何与游离细胞、细胞膜、下丘脑、胃肠道或稳定的神经胶质细胞株NG_{108-15}特异结合的$^{125}I-CART_{61-102}$。其原因可能是在正常细胞和肿瘤细胞中存在不同的CART受体或组织中的结合位点密度比培养细胞低得多。同时，通过小鼠喂养试验，Lenka证实$CART_{61-102}$碘化后生物活性的部分保守性，1μg/只鼠剂量的I_2-CART_{61-102}与0.5μg/鼠的$CART_{61-102}$相同程度地降低食物摄取量。与^{125}I – CART$_{61-102}$和 PC12 细胞的特异性结合一起证明了^{125}I 放射性标记并没有在很大程度上改变 CART 的药理性能。

  总之，无论是未分化的 PC12 细胞还是分化的 PC12 细胞，其上^{125}I – CART$_{61-102}$结合位点的发现以及二碘化 CART$_{61-102}$生物活性保留可用于进一步研究潜在的 CART 受体分子的特征。

第二章

研究意义、内容和技术路线

第一节　研究意义

在生产实践中，为了提高优良品种的繁殖率，通常采用卵母细胞体外受精或外源激素超排途径。但这些方法技术含量要求高，生产成本高，成活率低，在养殖户中推广普及难度大。探讨一种能在卵巢内解决卵泡闭锁问题的途径，达到自然超排效果，结合胚胎移植技术，为成倍地提高畜群繁殖效率，加快优良种畜的扩繁速度提供先进的繁育技术支持。

CART是一种下丘脑神经肽，在动物体内执行多种功能，包括抑制摄食、减轻体重、调节能量代谢平衡、调节机体应激反应、调控下丘脑—垂体—性腺轴和下丘脑—垂体—肾上腺轴、调节自主神经和感觉传导、产生或促进精神兴奋药物引起的动物条件性位置偏爱、参与疼痛传导及镇痛，此外，对人的研究发现CART还具有神经营养功能，促进培养的神经元发育及存活，以及CART与Ⅱ型糖尿病胰岛素抵抗有关。近来发现CART在牛卵巢卵泡中表达，是影响卵泡发育的局部负调控因子，与卵泡闭锁密切相关。

本研究旨在通过转录组深度测序、蛋白质组学研究等技术探讨单胎家畜牛在卵泡发育过程中，与 CART 相互作用的蛋白，进而通过同源建模、分子对接技术筛选 CART 的受体，为进一步分析单胎家畜卵泡闭锁提供理论依据，进而为提高单胎家畜的排卵率，解决优良种畜扩繁等提供技术支持；有助于进一步揭示人类相关疾病的发生发展过程，对新药物的研制开发和治疗人类疾病具有重大意义，并有望为糖尿病治疗提供新的思路同时，也形成了一条研究神经肽受体的成熟通路。

第二节 研究内容

1. B 超声波监测牛卵巢卵泡发育，采集 ODF1 和 ODF2；分离颗粒细胞并进行转录组高通量测序，并对测序结果进行分析，利用 CELLO V2.5 对差异表达基因进行亚细胞定位。

2. 提取 GCs 总蛋白，用兔抗 CART 一抗沉淀蛋白复合体；经 LC - ESI - Q - TOF 质谱仪分析蛋白复合体，建立 CART 相互作用蛋白网络图，并对获得的蛋白进行亚细胞定位。

3. 应用 HMMTOP V2.0 从膜蛋白中筛选 G 蛋白偶联受体；通过 SWISS - MODEL 和 MODELLER 同源建模技术，建立 CART 及 G 蛋白偶联受体的三维分子坐标。

4. 通过 ZDOCK 分子对接技术获得 CART 与 G 蛋白偶联受体复合体的三维模型，通过评分函数获得评分最高的 3~4 个 G 蛋白偶联受体作为 CART 的受体。

5. 通过黄体形态观察、E_2 和 P 测定获得牛发情周期内第一卵泡波 DF 和 SF；分别分离颗粒细胞、提取总 RNA，利用 qRT - PCR 技术检测 CART 及其受体的表达量差异。

第三节 技术路线

　　研究中通过采集牛发情期第一卵泡波 ODF1 和 ODF2 卵泡，通过对 2 个转录组高通量测序，从中筛选高差异表达基因；通过免疫共沉淀及 LC-ESI-Q-TOF 质谱分析试验及 Cytoscape V3.1.0 分析，构建 CART 蛋白相互作用网络图；在此基础上，应用多项生物信息学技术，从高差异表达基因和 CART 相互作用蛋白中筛选膜蛋白，并通过跨膜区预测 G 蛋白偶联受体；综合应用 SWISS-MODEL 和 MODELLER 同源建模技术对 CART 和预测的 G 蛋白偶联受体进行分子建模；经 ZDOCK V3.0.2 将 CART 分别与预测的 G 蛋白偶联受体逐一进行分子对接，ZDOCK 通过静电能量、几何匹配的运算途径求卷积的方法来求解，应用 FFT 方法加速，使用基于几何匹配的打分方式 PSC（Pairwise shape complementarity）和基于原子配对统计的打分方式 ACE（Atomic contact energy）进行评价，最终将评分值最高的三个 G 蛋白偶联受体作为 CART 的候选受体；通过实时荧光定量 PCR 技术对 3 个候选受体在牛卵泡颗粒细胞的表达情况分析。具体研究技术路线如图 2-1 所示。

第二章 研究意义、内容和技术路线

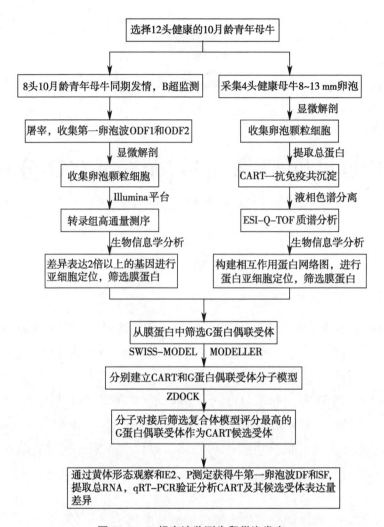

图 2-1　B 超声波监测牛卵巢泡发育

第三章

牛卵巢卵泡高通量测序与 GO 分析

第一节 概述

Illumina 平台测序时，使用了 Solexa 测序芯片，它含有 160 多万个由光纤组成的孔，孔中载有化学发光反应所需的各种酶和底物。测序开始时，放置在 4 个单独试剂瓶里的 4 种碱基，依照 T、A、C、G 顺序依次循环进入 PTP 板，每次只进入 1 个碱基。如果发生碱基配对，就会释放 1 个焦磷酸。这个焦磷酸在各种酶的作用下，经过 1 个合成反应和 1 个化学发光反应，最终将荧光素氧化成氧化荧光素，同时释放出光信号。此反应释放出的光信号实时被仪器高灵敏度 CCD 捕获。有 1 个碱基和测序模板进行配对，就会捕获到一分子的光信号；由此一一对应，就可以准确、快速地确定待测模板的碱基序列。

本研究通过 B 超声波监测技术采集牛发情周期第一卵泡波中最大卵泡 ODF1（The largest follicle at onset of deviation，发情后的第 4～5d）和第二大卵泡 ODF2（The second largest follicle at onset of deviation），分别分离颗粒细胞并提取总 RNA，构建 RNA 文库并通过 Illumina 平台对 ODF1 和 ODF2 测序，分别获得 44 189 827 和 43 826 914 个原始"reads"；经过滤、装配、数据库比对，最终获到 35 325 个已经注释的

转录本；对所获得的 35 325 个基因进行 Gene Ontology（GO）功能显著性富集分析。该分析首先把所有表达基因向 GO 数据库（http：//www.geneontology.org/）的各个 term 映射，计算每个 term 的基因数目，然后应用超几何检验，找出与整个基因组背景相比显著富集的 GO 条目。GO 分类系统结果显示，所获得的 35 325 个基因共分为三大类 51 组：其中，生物学过程占 42.43%，细胞组分占 26.43%，分子功能占 31.14%；设定条件 ODF1 RPKM/ODF2 RPKM > 2，FDR 校正后的 $P <0.05$，在整个文库中共获得 1 331 个差异表达基因，经 CELLO V2.5 进行亚细胞定位，细胞核蛋白占 44.38%，细胞质蛋白占 13.77%，叶绿体蛋白 1.59%，胞外蛋白 17.42%，线粒体蛋白 9.65%，内质网蛋白占 0.06%，膜蛋白占 12.71%，高尔基体蛋白占 0.18%，溶酶体蛋白占 0.18%，过氧化物酶蛋白占 0.06%，其中，膜蛋白有 188 个。通过膜蛋白的筛选，进一步缩小 CART 受体的筛选范围。

第二节 材料与方法

一、试验动物及样品采集

（一）卵泡采集

选取 8 头 10 月龄青年母牛（海福特杂交后代），注射 PGF2α(前列腺素 F2α) 使它们达到同期发情，分别进行为期 11d 的饲养管理，在此期间每天用 B 超仪检测并记录卵泡的生长状况。在第一卵泡波的卵泡发育偏离期启动时，屠宰并摘除卵巢，用眼科剪剪下出现偏差时的最大卵泡 ODF1 和第二大卵泡 ODF2，放入灭菌的 DPBS 中准备分离颗粒细胞。

（二）颗粒细胞分离

在超净台内用 1ml 注射器将卵泡液抽出后置于灭菌的 EP 管中，置于 -20℃ 冰箱中保存，以备后续雌激素、孕激素的测定。之后用眼科剪将卵泡剪开一个小口，放到盛有 DPBS 的表面皿上，用细胞刮刀轻轻刮取附于卵泡内膜的颗粒细胞并在 DPBS 中清洗，用移液枪吸取含有颗粒细胞的 DPBS 混合液于灭菌 EP 管中离心，弃上清液并加入 1ml RNAiso Plus，投入液氮预冷数秒后转存于入 -80℃ 冰箱中保存，以备后续 RNA 的提取。

二、主要试剂及耗材

Solexa 测序芯片、TruSeq RNA Sample Prep Kit、QiaQuick PCR kit、TruSeq SBS kit、裂解缓冲液（Illumina，美国）、PrimeScript© RT reagent Kit With gDNA Eraser、Rnase-free H_2O、RNAiso Plus（Takara，大连），固相 RNase 清除剂（Andybio，美国），液氮、异丙醇、氯仿、无水乙醇、无 RNase 枪头、1.5ml 无 RNase EP 管、PE 手套和口罩等。

三、试验方法

（一）颗粒细胞 Total RNA 的提取

从 -80℃ 冰箱中取出含有颗粒细胞和 RNAiso Plus 的混合溶液，不停吹打直至细胞充分裂解，固形物消失，室温静置 5min，Trizol 法提取 Total RNA，-80℃ 保存。

（二）Total RNA 完整性和浓度的检测

将 Total RNA 与 Loading Buffer 按 5:1 混合后加样于配制好的 1% 的琼脂糖凝胶，电泳 10min 左右检测 Total RNA 的完整性。利用核酸蛋白测定仪在 OD260、OD280 下检测 Total RNA 的浓度和纯度，要求 OD260/OD280 在 1.8~2.0 间。

（三）牛卵泡颗粒细胞 RNA 文库构建并测序

取检验合格的 RNA 样品 6μl，mRNA 经带有 Oligo（dT）的磁珠富集后，再向其中加入裂解缓冲液，使其裂解为大小为 200nt 左右的短片段，然后以这些短片段为模板，加入随机引物六聚体合成 cDNA 第一链。合成的 cDNA 第一链经过加入一系列的缓冲液、RNase H、dNTPs 和 DNA 经聚合酶 I 合成 cDNA 第二链，再经过 QiaQuick PCR kit 纯化、EB 缓冲液的洗脱、末端修复、加 PolyA 尾巴、加 5'和 3'接头，最后通过琼脂糖凝胶电泳、回收目的片段并通过 PCR 扩增，从而完成整个 cDNA 文库构建并使用 Illumina HiSeq 2000 测序平台进行测序。

四、生物信息学分析

（一）重新装配与功能注释

Illumina 测序得到的原始"reads"经过去除接头、低质量的"reads"以及空"reads"后获得的读段为"clean reads"，然后使用 CLC 基因组学工作平台（CLC bio，Aarhus，Denmark）把这些"clean reads"映射定位到牛 RefSeq 数据库中。通过计算 RPKM（Reads per kilobase per million reads）值来评估基因的表达量，RPKM = 10^9C/（NL），其中，C 代表唯一映射到 RefSeq 数据库中某基因的数目，N 代表唯一映射定位到参考基因组中的总读段数，L 代表某基因的碱基数。计算出所有基因的 RPKM 值。

（二）亚细胞定位

通过 CELLO V2.5 Subcellular Localization Prediction 数据库对 1 331 个基因进行亚细胞定位。首先通过 NCBI 中的 FASTA 搜索引擎分别找出每一个基因的全序列，物种选择牛；然后将该基因的 FASTA 序列输入 CELLO V2.5 对话框，并对结果进行统计。

第三节 结果与分析

一、牛卵泡颗粒细胞 total RNA 提取结果

分别对符合要求的牛卵巢 ODF1 和 ODF2 颗粒细胞进行总 RNA 的提取，样品经微量核酸蛋白测定仪纯度检测后，OD260/OD280 均在 1.8 ~ 2.1，浓度均高于 600ng/μl，纯度和浓度均符合后续试验要求。同时为了全面检测所提取颗粒细胞总 RNA 的完整性，采用 1% 琼脂糖凝胶电泳检测，结果发现两组样本的总 RNA 纯度较高且完整性好，可直接用于下一步深度测序（图 3-1）。

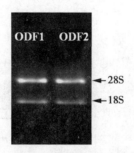

图 3-1　总 RNA 电泳图

Figure 3-1　Gel electrophoresis figure for total RNA

二、Illumina 测序与序列重新装配及 GO 功能富集分析

通过 Illumina 平台对 ODF1 和 ODF2 测序，分别得到 44 189 827 和 43 826 914 个原始"reads"。经过去除接头、低质量"reads"以及空"reads"等一系列过滤过程后，分别产生了 44 082 301 和 43 708 132 个"clean reads"。通过 Trinity 短片段装配软件把这些"clean reads"以一定长度的层叠群连接起来，从而形成"contigs"（图 3-2A）。然后这些"reads"被重新映射到拥有双末端"reads"的"contigs"中，这样就能

第三章
牛卵巢卵泡高通量测序与 GO 分析

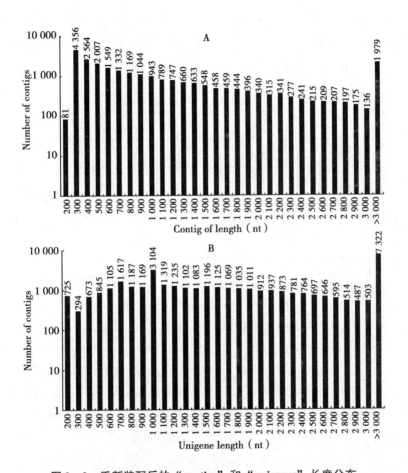

图 3-2 重新装配后的 "contigs" 和 "unigenes" 长度分布

Figure 3-2 Length distributions of the de novo assembly for contigs and unigenes.

注：X 轴的每一个数值代表覆盖了 100bp 的序列长度范围，例如："200" 代表一个序列长度范围 [200, 300)。Note: Each number in the x-axis indicates a region of sequence length covering 100 bp. for example, "200" represents a region of sequence length [200, 300).

够从相同的转录本中检测 "contigs" 以及它们之间的距离。最后把这些 "contigs" 连接起来获得共有序列，这样的序列被称为 "unigenes"（图 3-2B），之后通过 BLASTX 将这些 "Unigene" 序列比对到 nr, Swiss-Prot and KEGG 等数据库中（e < 0.00001），获得 35 325 经过注释的转录

31

本。通过 GO 分类系统对 ODF1 和 ODF2 中获得的 35 325 个转录本进行 GO 分类,共分为三大类 51 组。其中,生物学过程占 42.43%,细胞组分占 26.43%,分子功能占 31.14%(图 3-3)。

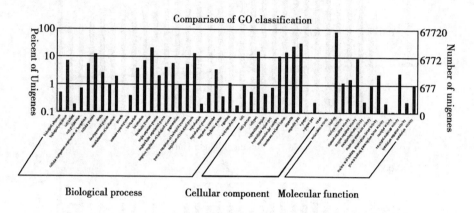

图 3-3　"unigenes" 的 GO 分类

Figure 3-3　GO classification of the unigenes

注:左边纵坐标表示在某一特定功能分类中"unigenes"所占的比例,

右边纵坐标表示在某一分类中"unigenes"的基因数目。

Note: The right y - axis shows the number of genes in a category, while the left y - axis indicates the percentage of a specific category of genes in that main category.

三、差异表达基因筛选及产物亚细胞定位

为了从表达水平上鉴定出 ODF1 和 ODF2 中存在显著差异的基因,本研究参照 Audic 等研究发现的算法,从 ODF1 和 ODF2 两个文库中得到这些差异表达基因。设定条件为:ODF1 RPKM/ODF2 RPKM > 2,FDR 校正后的 $P < 0.05$,在整个文库中共获得 1 331 个差异表达基因。

通过大量的研究,虽然 CART 受体的分子特性尚未明确,但 CART 的生物学作用是由受体介导的。CART - GFP 融合蛋白与 Hep G2 细胞和分离的下丘脑细胞的结合已有报道;CART 与 AtT20 细胞、PC12 细胞和

原代培养大鼠伏隔核细胞的特定饱和结合也已证明，当 GTP 存在时 CART 结合会减少，但不受 ATP 类似物存在的影响；此外，在研究 CART 在颗粒细胞、AtT20 细胞和海马神经元的作用时，当用和 CART 受体信号有关的 o/i 子类抑制性 G 蛋白的抑制剂预处理细胞时，细胞上的 Go/i 信号终止。所有的研究证据表明 CART 的生物学作用可能是通过 G 蛋白偶联受体介导的。Joseph 等（2013）在研究牛卵泡颗粒细胞 E_2 产生过程中 CART 信号通路时，也证明 CART 生物学作用的实现是由其受体介导的，其受体可能是 G 蛋白偶联受体。而 G 蛋白偶联受体是一类膜蛋白受体的统称。

通过对深度测序结果进行分析，其中，ODF1/ODF2 大于 2 的差异表达基因共 1 331 个，经 CELLO V2.5 Subcellular Localization Prediction 数据库分析进行亚细胞定位，结果显示（图 3-4）：定位于细胞核的蛋白占 44.38%，细胞质蛋白占 13.77%，叶绿体蛋白 1.59%，胞外蛋白 17.42%，线粒体蛋白 9.65%，内质网蛋白占 0.06%，膜蛋白占 12.71%，高尔基体蛋白占 0.18%，溶酶体蛋白占 0.18%，过氧化物酶蛋白占 0.06%。其中，膜蛋白有 188 个。

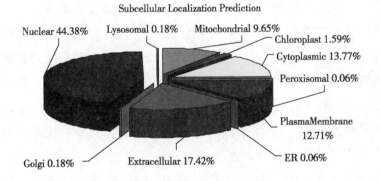

图 3-4 亚细胞定位预测

Figure 3-4 Subcellular localization prediction

第四节 讨论

转录组是某一组织或细胞在特定发育阶段或功能状态下，所有RNA转录的集合。对转录组进行研究，能够从组织或细胞的整体水平研究各基因功能及其基因结构，是研究特定生物学过程分子机理的一种高效手段。

Illumina 公司的 HiSeq 2000 测序仪是目前使用最多高通量测序平台，核心技术是"DNA cluster"和"Reversible terminator（可逆性末端终结）"，运用边合成边测序的机理，工作流程如下。

①构建测序文库。将 DNA 链随机打断并 DNA 链两端加上接头；②锚定桥连。每个 DNA 接头与测序通道上的接头引物随机结合并进行 PCR 扩增；③DNA 簇的产生。通过变性和桥式扩增循环获得大量成簇的待测 DNA 片段；④单个碱基的延伸测序。将标记 dNTP、引物和聚合酶加入测序通道，启动测序循环。通过激光激发，从各个测序通道的测序簇中产生荧光，然后利用捕获荧光的颜色获得测序簇碱基。

新一代测序技术与 DNA 芯片相比较，基于杂交技术的微阵列芯片只能用于检测已知序列，且灵敏度有限，难以检测丰度较低的目标，更无法捕捉基因表达水平的细微变化，而这正是研究特定阶段生物反应所必需的。新一代测序技术通过对基因表达的定量分析，可以更准确地检测组织或细胞 RNA 表达水平。Marioni 等（2008）对新一代测序和芯片技术在检测差异表达基因方面作了比较研究，相同的肝脏和肾脏 RNA 样品经基因芯片（Affymetrix 公司）和 Illumina 测序平台测序，在相同 FDR（False discovery rate）的情况下，新一代测序技术比基因芯片多检测出 30% 的差异表达基因。Carlos 等研究表明，Illumina 技术具有高度重复性，技术误差相对较小。

到目前为止，人们利用传统的 Sanger 测序技术与微阵列技术已经

第三章 牛卵巢卵泡高通量测序与 GO 分析

对牛卵泡发育做过很多研究，但是绝大多数报道都是基于超声成像与细胞体外培养。为了能够找到更多基因在卵泡发育过程所行使的分子生物学功能，我们采用 Illumina 测序平台对牛卵泡颗粒细胞进行深度测序，并对所获得的基因表达水平进行评估，为以后进一步研究它们的功能奠定基础。经过研究，获得近 9 亿转录组序列"reads"以及 35 325 个基因在牛卵泡颗粒细胞中表达。

小结

1. 通过对牛卵泡进行深度测序，共发现 35 325 个基因在牛卵泡 ODF1 和 ODF2 中表达。

2. 对 35 325 个基因进行 GO 分类，共分为三大类 51 组。其中，生物学过程占 42.43%，细胞组分占 26.43%，分子功能占 31.14%。

3. 获得 ODF1/ODF2 大于 2 倍的差异表达基因 1 331 个，亚细胞定位结果显示：细胞核蛋白占 44.38%，细胞质蛋白占 13.77%，叶绿体蛋白 1.59%，胞外蛋白 17.42%，线粒体蛋白 9.65%，内质网蛋白占 0.06%，膜蛋白占 12.71%，高尔基体蛋白占 0.18%，溶酶体蛋白占 0.18%，过氧化物酶蛋白占 0.06%。其中，膜蛋白有 188 个。

第四章

牛卵泡颗粒细胞 CART 蛋白质组学研究

第一节 牛卵泡颗粒细胞总蛋白的提取

一、概述

生命体的功能执行者是基因的表达产物蛋白质。DNA – mRNA – 蛋白质，存在3个层次的调控，分别是转录水平调控、翻译水平调控和翻译后水平调控，从 mRNA 水平进行的研究仅包括转录水平调控，并不能全面代表蛋白质表达水平，基因产物是否表达以及何时被翻译表达、翻译产物的相对浓度、翻译后的修饰程度等问题并不能通过研究基因组得到解答。蛋白质是生理功能的执行者，是生命现象的直接体现者。对蛋白质结构和功能的研究将直接阐明生物体在生理或病理条件下的变化机制。

本试验中，母牛屠宰后采集双侧健康卵巢，选取 8～13mm 卵泡，分离颗粒细胞，Western 及 IP 细胞裂解液进行颗粒细胞总蛋白的提取。

二、材料与方法

（一）试验动物及其组织

本试验材料来自北京市大兴区魏善庄镇金维福仁肉牛屠宰场。选取

4头母牛（海福特杂交后代），屠宰后采集双侧卵巢，编号后投入冰盒中制备好的杜氏磷酸缓冲液（DPBS）中，迅速在实验室进行样品处理，包括观察黄体的形态特征、分离大小卵泡GCs，并迅速投入液氮罐。

（二）主要试剂及耗材

Western及IP细胞裂解液（Beyotime，上海），蛋白酶抑制剂（Roche，瑞士），异丙醇、正己烷、无水乙醇、甲醛等均为国产分析纯级试剂，液氮、EP管、橡胶手套、口罩、眼科剪和镊子等。

（三）试验方法

1. 颗粒细胞的分离

牛屠宰后，采集双侧卵巢上的健康大卵泡，要求卵泡大小在8~13mm，应是透明的或呈黄色，这样的卵泡有可能是卵巢上的DF；避免使用充血的、大于13mm的或者呈黑色的卵泡，这样的卵泡可能是上一发情周期内卵泡波的退化卵泡；也不能使用存在20mm或更大卵泡的卵巢，这很可能是卵巢上出现囊肿，会影响其他卵泡的正常发育。

2. 颗粒细胞总蛋白的提取

GCs总蛋白的提取参照Demasi方法并适当进行调整，将EP管中的优势卵泡GCs用预冷的PBS液冲洗2次，细胞悬液转移到灭菌的离心管中，1 000g/min，4℃离心10min，弃上清液；低渗溶液重悬，1 000g/min，4℃离心5min，弃上清液；低渗溶液重悬沉淀，冰上溶胀35~40min，3 000g/min，4℃离心15min，弃上清液；逐滴加入Western及IP细胞裂解液，冰上孵育30min；20 000g/min，4℃离心10min，取上清液，此即为颗粒细胞总蛋白，装于EP管-80℃保存备用。

3. 总蛋白浓度测定

利用核酸蛋白测定仪检测总蛋白的浓度和纯度。蛋白质在280nm紫外吸收值大于260nm的吸收值，通常，纯蛋白质的光吸收比值在1.8~2.0间说明总蛋白纯度较高。

三、结果与分析

经核酸蛋白测定仪检测总蛋白的浓度和纯度。结果显示：OD280/OD260 = 1.85，说明总蛋白纯度较高；总蛋白浓度为 1.78μg/μl。由于试验中提取总蛋白时，GCs 数量较少，故总蛋白浓度较低，但可以用于免疫共沉淀（Co – Inununoprecipitation，Co – IP）试验。

四、讨论

生物膜是几乎是所有生命存在的必要条件，膜蛋白几乎占到所有蛋白总量的1/3。从低等的原核生物到高等的真核生物，膜的结构和功能也越来越复杂。完整的膜结构一般由脂质双分子层和嵌入或覆盖在其表面的脂类分子和蛋白分子组成。完整的膜蛋白组在信号传导和物质、能量传递上发挥重要作用。因此，膜蛋白通常是医药领域研究的重要靶标。由于膜蛋白组分具有极难溶解、丰度偏低、成分复杂、蛋白修饰多样、动态变化明显等特点，使得膜蛋白质的研究面临重大挑战。膜蛋白质研究技术通常需要具备对整体蛋白进行精确分类，可靠衡量蛋白表达水平的标准，准确对蛋白丰度改变，蛋白之间互作改变，蛋白位置改变等进行定性、定量测定。目前，在膜蛋白的溶解试剂中常用到强碱盐水，通常加入适当的氯仿/甲醇、Triton X – 114 等辅助试剂。另外，有些方法利用尿素、盐酸、Brij – 58、RapiGest 等替代有机溶剂。

本研究中使用 Western 及 IP 细胞裂解液（Beyotime，上海）将颗粒细胞重悬，反复冻融。其原理是通过细胞重悬和反复冻融破坏细胞膜结构，以便后续的提取。提取过程中应尽可能保持蛋白原来的状态，不丢失其天然活性。

第二节 免疫共沉淀及质谱分析CART相互作用蛋白

一、概述

蛋白质在执行生理功能时表现出多样性和动态性，而不像基因组那样基本不变，如蛋白质的修饰加工、转运定位、结构形成、代谢等均无法从基因组水平获得，需要在整体、动态、网络的水平上对蛋白质进行研究。蛋白质组学的研究内容包括蛋白质的定量分析、定性鉴定、蛋白质在细胞内定位及相互作用蛋白的研究等，最终揭示蛋白质功能。蛋白质分离技术、鉴定技术、相互作用蛋白分析技术及生物信息学技术是蛋白质组学研究主要依赖的四大技术。

蛋白质组学研究能否顺利进行，在很大程度上取决于研究技术和方法水平的高低。对蛋白质进行研究要比对基因复杂得多，如氨基酸残基种类远多于核苷酸残基（20/4），且蛋白质存在复杂的翻译后修饰，对蛋白质组的研究是从细胞水平上对蛋白质进行大规模的分离和分析。因此，发展高通量、高灵敏度、高准确性的研究技术平台是蛋白质组学研究中的长期任务。

蛋白质相互作用的研究技术包括从原子水平、分子水平和细胞水平进行探讨。相互作用蛋白质组学的核心是从分子水平上去研究蛋白质间直接的相互作用，研究方向集中在两两蛋白质间相互作用和蛋白复合物的相互作用研究上。高通量大规模研究蛋白质相互作用的方法目前主要是蛋白质芯片(Protein chip)、串联亲和纯化(Tandem affinity purificatinn, TAP)和酵母双杂交(Yeast two-hybrid System)；小规模的蛋白质相互作用分析包括化学交联、Far-western、GST pull-down 和 Co-IP 等。

目前，用于蛋白质质谱分析的软电离技术主要有：MALDI-MS、大气压电离质谱（Atmospheric pressure Ionization-Mass spectrometry, API-MS)、ESI-MS、快原子轰击质谱（Fast atom bombardment-Mass spectrometry，FAB－MS）等。随着电离技术的更新与完善，质谱用于蛋白质的检测可达到 fmol 级，可测定分子量最高达 100 000Da。目前，生物质谱技术已成为测定生物大分子，尤其是蛋白质和多肽分子量和一级结构的重要工具，也是生命科学领域蛋白质组研究中不可或缺的关键技术。

本试验选用兔抗 CART 一抗沉淀 CART 及其相互作用蛋白，经 Protein G－Agarose 磁珠吸附分离，多次洗涤去除盐分和杂蛋白，Western 及 IP 细胞裂解液洗脱蛋白复合体并酶解。经 LC－ESI－Q－TOF 质谱分析及数据库比对，共鉴定蛋白质组分 111 个。

二、材料与方法

（一）试验材料

试验材料为保存的牛卵泡颗粒细胞总蛋白（具体方法见第一节牛卵泡颗粒细胞总蛋白的提取）。

（二）主要试剂及耗材

兔抗 CART 一抗（Phoenix Pharmaceuticals，美国），Western 及 IP 细胞裂解液（Beyotime，上海），酶解试剂盒（Pierce，美国），Protein G－Agarose 磁珠、蛋白酶抑制剂（Roche，瑞士），色谱级甲醇、乙腈、氯仿、异丙醇、正己烷、无水乙醇、甲醛、NaCl、KCl、Na_2HPO_4、KH_2PO_4 等均为国产分析纯级试剂，液氮、EP 管、橡胶手套、口罩、眼科剪和镊子等。

（三）试验方法

1. 蛋白非特异性结合的去除

将柱子放在离心管上，按每 1mg 初蛋白加 80μl Control Agarose Resin 过柱；1 000g 离心 1min，弃下清液；100μl 1×Coupling Buffer 洗柱，并重复 1 次；2mg 初蛋白过柱，4℃孵育 1h；1 000g 离心 1min，收集下

清，此即为预处理的总蛋白。

2. CART 抗体免疫沉淀

取全部总蛋白样品并加入等体积的 Lysis Buffer；取 200μl Protein G – Agarose 磁珠，加 PBS 于 1ml EP 管中清洗磁珠，4℃条件下 3 000rpm 离心 3min，弃上清，重复 3 次；用 Lysis Buffer 清洗 1 次，4℃条件下 3 000rpm 离心 3min，使用前弃上清并与预处理蛋白混合；加 30μl 磁珠于总蛋白样品中，4℃摇床孵育 1h；4℃条件下 3 000rpm 离心 3min；取上清置于一无菌 EP 管中，加 2 μl 稀释过的 CART 抗体，4℃摇床孵育过夜；加入 50μl 磁珠，4℃摇床孵育 2~3h；4℃条件下 3 000rpm 离心 3min；磁珠分别用 500μl PBS 清洗 3 次，4℃条件下 3 000rpm 离心 3min，弃上清液；加入 10μl 5 × Sample Buffer 和 20μl Lysis Buffer，– 20℃保存备用。

3. CART 免疫共沉淀复合物的洗脱及酶解

将 Co – IP 磁珠转移至一无菌 EP 管中，用于收集洗脱液；加入 10μl Elution Buffer 洗涤吸附有 CART 免疫复合物的磁珠，1 000g 离心 1min，收集洗脱液；柱子中加 50μl Elution Buffer，室温孵育 5min，1 000g 离心 1min；收集 2 次洗脱液，即为 CART Co – IP 复合物，核酸蛋白检测仪测定蛋白浓度；加入 0.25% 胰酶消化液对 CART Co – IP 复合物进行酶解。

4. 高效液相色谱（HPLC）分离肽段

（1）纳流泵：流速控制为 0.3μl/min；将流动相溶剂 A（0.1% 甲酸 + 98.9% ddH$_2$O + 1% 乙腈）装入 HPLC – Chip 富集柱 1.5min，从富集柱反冲后，样品被装到 RP 分析柱，2min 内用梯度至 15% 的溶剂 B（0.1% 甲酸 + 99.9% 乙腈水）梯度洗脱。

（2）毛细管泵：流动相为 99% ddH$_2$O + 1% 乙腈，流速 4.0μl/min。

（3）进样量：8.0μl。

5. 质谱分析

质谱仪使用 LC – ESI – Q – TOF，所有测定均在正离子模式下进行，

牛卵泡可卡因－
苯丙胺调节转录肽（CART）受体的筛选

获取的 MS/MS 通过 MassLynx（Micromass, Manchester, UK）软件转化为 PKL 文件。MASCOT v2.2 软件（www.matrixscience.com）对获得的 PKL 文件进行数据库搜索，参数设定为：IPI（The international protein index）Mammalia database（Littp：//www.ncbi.nl－m.nih.gov/）；固定修饰选择 Carbamoylmethylation（C），可变修饰为 Oxidation（M），Trypsin 切割，允许最大未酶切位点数为 1，肽段质量容差（Peptide mass tolerance）为 ±50 ppm，碎片离子质量容差（Fragment mass tolerance）为 ±0.05 Da。肽段得分超过阈值的视为鉴定肽段，其对应的蛋白质为鉴定蛋白质，具有置信度大于 95%，阈值标准由制造商制定。对于由两个或两个以上肽段同时匹配的、得分超过 20 的蛋白质，认为该蛋白质鉴定准确，不需要人工检查；得分在 20 分以下的需系统的人工验证，进而确定结果的取舍；只有单一肽段匹配的蛋白质，其离子得分必须超过 20 分，并通过人工检验确定结果取舍。

三、结果与分析

（一）LC－ESI－Q－TOF 质谱鉴定结果

本研究中将 Co－IP 分离的蛋白复合体样品经胰蛋白酶酶解得到多肽混合物，进入液相色谱（Liquid chromatogram, LC）分离，经一级质谱（MS）检测获得分离后多肽分子离子信息，从中筛选出的若干分子离子，每一个分子离子经 CID（碰撞诱导解离）获得二级质谱（MS/MS）数据，数据库查询得到该多肽所属蛋白质序列，这样就完成了蛋白质的定性。通过从头测序（De Novo）技术（也称鸟枪法），检测结果显示：一级质谱检测获得 2 522 个多肽分子离子信息，这些多肽分子离子经 CID 获得 7 018 个二级质谱碎片离子信息。

由于每一种搜索工具本身的不完善，同时蛋白质数据库在构建中存在不同缺陷，造成数据库搜索结果可能存在大量假阳性结果，导致高误判率。搜索目标－诱饵库（Target－decoy database）是一种普遍使用的评估结果假阳性的方法。图4-1为诱饵库（粉色）与目标库（蓝色）混合构成混合数据库，由图4-1（a）可以看出，随着峰值肽得分增加，PSM数呈下降趋势；由图4-1（b）可以看出，随着峰值肽得分的增加，假阳性率明显降低，母离子质量误差集中在$1\times10^{-7}\sim3\times10^{-7}$，混合数据库分界——峰值肽得分（-10lgP）控制在≥23.4（表4-1）；图4-2为经PSM过滤后母离子和多肽分子离子m/z的质量误差均达到百万分之一级。图4-3为从头测序肽段置信度分布，图4-3（a）显示诱饵库（粉色）与目标库（蓝色）重合面积较图4-3（b）图要小，与目标库频谱匹配肽段非重合区置信度均在50%以上；图4-3（b）为从头测序单肽段数据库赋值的置信度分布，目标库频谱匹配肽段非重合区置信度均在79%以上。质谱分析数据在Mascot搜索结果的基础上，采用诱饵蛋白质库搜索结果对Mascot搜索结果进行假阳性率（False discovery rate，FDR）估计。结果显示（图4-4）：当假阳性率控制在5.0%时，共有2 921个肽段频谱与数据库相匹配；在本研究中，参照Balgley等使用1.0%谱图假阳性率对应的E值作为阈值来过滤Mascot的搜库结果，从而确定肽段频谱与数据库匹配数为2 566个。

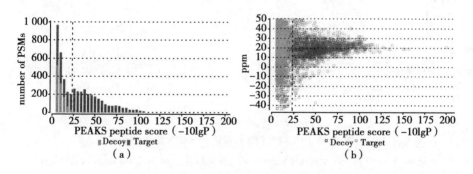

图4-1 肽段频谱匹配（PSM）得分分布

Figure 4-1 Peptide-spectrum matches (PSM) score distribution

注：（a）图为峰值肽的得分分布；（b）图为峰值肽得分的散点图与母离子质量误差。

Note：（a）Distribution of PEAKS peptide score；（b）Scatterplot of PEAKS peptide score versus precursor mass error.

牛卵泡可卡因－苯丙胺调节转录肽（CART）受体的筛选

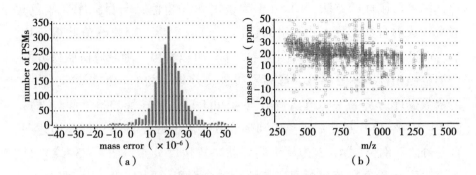

图4－2　PSM过滤结果的母离子质量误差

Figure 4－2　Precursor mass error of peptide－spectrum matches（PSM）in filtered result

注：（a）图为母离子质量误差在ppm级的分布；

（b）图为母离子质荷比与质量误差在ppm级的散点图

Note：（a）Distribution of precursor mass error in ppm；

（b）Scatterplot of precursor m/z versus precursor mass error in ppm.

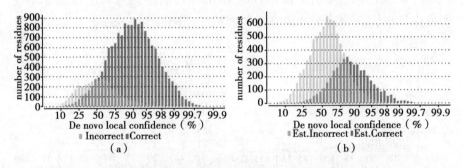

图4－3　从头测序结果确认——肽段置信分布

Figure 4－3　De novo result validation. Distribution of residue local confidence

注：（a）图为通过数据库肽赋值确认从头测序肽段；

（b）图为从头测序单肽段的数据库肽赋值

Note：（a）Residues in de novo sequences validated by confident database peptide assignment；（b）Residues in "de novo only" sequences.

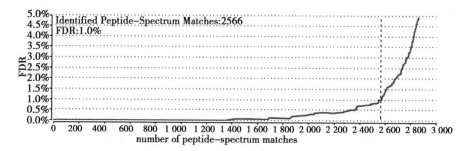

图 4-4 假阳性率（FDR）曲线

Figure 4-4 False discovery rate (FDR) curve

注：X 轴为 PSM 数，Y 轴为对应的 FDR。

Note: X axis is the number of PSM being kept. Y axis is the corresponding FDR.

（二）测序蛋白质组研发

从头测序蛋白质组研究策略中，卵巢卵泡 GCs 总蛋白经 Co-IP 形成的蛋白复合体未经 2-DE 分离纯化便被胰蛋白酶切成肽段，蛋白质与肽段之间的相关联信息可能发生丢失，增大了肽段推导蛋白质的困难。目前，简约原则是国际上蛋白质组装的通用原则，在蛋白质推导中被广泛应用。本研究使用 Mascot V2.2 作为数据搜索引擎，并基于 NCBI nr 数据库 (35 149 712 sequences；12 374 887 350 residues) 中哺乳动物（Mammalia, 2 144 738 sequences）数据库搜索策略展开研究，设定参数如表 4-1 所示，其中，选择峰值肽得分（-10lgP）≥23.4 的肽段进行蛋白质数据库匹配，匹配得分≥20，单肽段匹配数≥1，匹配置信度≥50%；显著性阈值（Significance threshold）$P<0.05$。数据库分析结果如表 4-2 所示，在 2 566 条频谱匹配肽段中，对应蛋白质组分 111 个。

表 4-1 数据库筛选参数

Table 4-1 Result filtration parameters

Parameters	Quantitative value
Peptide -10lgP	≥23.4
Protein -10lgP	≥20

(续表)

Parameters	Quantitative value
Proteins unique peptides	≥1
De novo ALC Score	≥50%

表 4-2 数据库筛选结果统计
Table 4-2 Statistics of filtered result

Filtered item	Result
Peptide-Spectrum Matches	2 566
Peptide Sequences	489
Protein Groups	111
Proteins	645
Proteins（#Unique Peptides）	140（>2）；83（=2）；422（=1）
FDR（Peptide-Spectrum Matches）	1.0%
FDR（Peptide Sequences）	4.1%
FDR（Protein）	0.0%
De Novo Only Spectra	1 109

非限制性 PTM 鉴定的质控方面基本参数设置如表 4-3 所示，半胱氨酸 C（Cysteine）上还原烷基化固定修饰 Carbamidomethylation（+57.02 Da）664 个，蛋氨酸 M（Methionine）上氧化可变修饰 Oxidation（+15.99 Da）24 个，丝氨酸/苏氨酸 ST 磷酸化 Phosphorylation（+79.97 Da）位点 14 个。

表 4-3 翻译后修饰（PTM）属性
Table 4-3 Post-translational modificatio (PTM) profile

Name	ΔMass	#PSM	Position
Carbamidomethyl	57.02	664	C
Oxidation	15.99	24	M
Phosphorylation	79.97	14	ST

四、讨论

生物质谱技术是近 20 年间发展起来的一项蛋白质组学分析技术。

其中 ESI 质谱通过纳电喷雾（nESI）大大提高了质谱分析对盐的耐受力和灵敏度，使极微量样品的分析成为可能；脉冲 ESI 则极大地减少了样品分析量，能在更低的流速下进行喷雾，显著提高质谱分析的灵敏度，MALDI 延时抽取（Delayed extraction）技术则极大地提高了质谱分析的分辨率；同时针对 MALDI 在氨基酸测序方面的不足，Chait（1993）发明并改进了一种"蛋白阶梯测序"法，即在 Edman 降解法基础上，生成大量去 N 端残基的肽段，利用 MALDI 测定其质量进而推测 N 端序列；Keough 等（1999）通过研究，利用肽段衍生化方法在肽段的 C 末端增加一个带负电荷的基团，这样 MALDI 产生的单电荷肽离子中就包含两个质子，一个位于 C 末端残基上，另一个则灵活性较强，可造成肽骨架多种方式的断裂，该方法显著增加了 MALDI 图谱的序列信息。

质谱法在检测和鉴定原核生物细菌方面应用较广，由于细菌所含蛋白质具有种属特征或与细菌的分类位置密切相关，可作为细菌分类依据，因此，细菌所含蛋白质标志物的检测已逐渐受到重视。Arnold 等（1999）利用 MALDI – TOF 质谱技术检测了 E. coli 核糖体蛋白和翻译后修饰产物，根据核糖体蛋白发生丢失或改变的情况，进而发现蛋白结构、遗传和功能方面的信息。Wilcox 等（2001）利用 MALDI – TOF 和 ESI – MS 技术分析了 E. coli 3 种耐药菌株的 56 种核糖体蛋白，通过测定质量变化，结果发现 S12、S5 和 L22 核糖体有突变发生，进而阐明了细菌耐药性的产生机制。其布勒哈斯（2009）利用 MALDI – TOF 技术对五种常见食源性致病菌的检测鉴定进行了系统研究，着重对不同基质、培养条件、预处理方法以及重复性分析对细菌蛋白图谱的影响进行了探讨，初步构建了食源性致病菌蛋白质图谱数据库。

Protein G – Agarose 磁珠 Co – IP 是利用抗原蛋白质和抗体的特异性结合以及细菌蛋白质的 "Protein A/G" 特异性地结合抗体（免疫球蛋白）Fc 片段的原理而开发的方法。目前多用 Protein A/G 预先结合在 Argarose beads 上，使之与含有抗原的溶液及抗体反应后，磁珠上的 Prorein A/G 就能达到吸附抗原的目的。通过低速离心，可以从含有目的抗原的溶液中将目的抗原与其他抗原分离。

Co-IP实验的操作步骤比较多，同时由于在非变性条件下进行实验，所以要得到一个完美的实验结果，不仅需要高质量的抗体，同时对免疫沉淀体系也需要有严格的控制指标。免疫沉淀实验从蛋白样品处理、抗体-agarose beads孵育、抗体-agarose beads复合物洗涤到最后的鉴定，每步都非常关键，需要严格控制实验流程中每个关键步骤的质量，才能达到最终实验目的。

LC-ESI-Q-TOF是以肽段鉴定单元作为鉴定结果，因此，对肽段鉴定的质量控制是实验中首要关注的问题。由于数据库搜索算法的局限性和质谱数据的复杂性，自动化软件从MS/MS到蛋白质鉴定面临许多问题。

首先，数据库搜索依赖于现有的公共数据库中肽段及蛋白质条目的正确性和完整性。例如，物种的基因组测序不完全，或蛋白质数据库不完整，这样就无法做出正确的鉴定；其次，搜索引擎打分算法的不足，也会造成肽段数据库匹配的遗漏，导致不能完全鉴定或错误鉴定的出现，产生假阳性/假阴性结果。此外，由于核酸多态性、点突变的存在，以及目标肽段含有未知修饰等，在数据库搜索算法或参数设置中未考虑这些因素，也会造成数据库无法给出真实的匹配结果；同时，随着数据库容量的增大，数据库搜索方法的速度和效率也会越来越低。Patterson在讨论质谱数据分析时指出：实验产生新数据的速度远远超出数据库的分析能力，对数据进行准确分析是蛋白质组学研究的核心问题。实验过程中获得的质谱数据质量高低取决于仪器类型、样品制备过程及实验人员操作技能等诸多因素，即使摒弃搜索引擎的影响，质谱鉴定率一般也低于30%，对于高精密度仪器数据，质谱鉴定率通常可以达到50%左右。

第三节 生物信息学分析

一、概述

生物信息学除了生物学研究外，重要的是信息科学中数据库的建立和更新，持久稳固地在一个稳定的存储介质中提供对海量数据的支持。生物信息学研究的数据主要在分子层面，主要的是使用了信息科学的技术，涉及了包括计算机技术、数学和统计学的方法论，用这些知识来组织和理解关于生物分子的相关信息。生物信息学的发展为人类生命科学研究提供了广阔的平台。它的发展不仅对相关学科基础研究起到了推动作用，而且必将对畜牧养殖、医学、食品等产业产生巨大的经济效益。

随着多物种基因组计划的实施，在后基因组时代对基因序列表达产物的功能分析将变得日益重要。蛋白质的亚细胞定位对蛋白质生物功能的发挥有着极其重要的作用，且蛋白质亚细胞定位信息可以暗示其生物学功能。通过亚细胞定位分离出各种亚细胞组分，极大地降低针对全细胞蛋白质组学研究的复杂性，同时，对蛋白质亚细胞定位信息的了解还可以促进对生物功能的认识，使得亚细胞定位成为蛋白质组学研究的基础。

本研究将质谱鉴定获得的 111 个蛋白质组分经 Cytoscape V3.1.0 分析，并构建了 CART 相互作用蛋白网络图；经 CELLO V2.5 Subcellular Localization Prediction 数据库对所获得的蛋白进行了亚细胞定位，为 CART 受体蛋白的筛选奠定基础。

二、研究方法

Cytoscape V3.1.0 对质谱分析所获得 111 个蛋白质组分构建蛋白相互作用网络图；通过 Cytoscape V3.1.0 对 CART 相互作用蛋白进行基因功能富集性分析，选择牛基因组数据库，超几何分布检验及本杰明－哈氏假阳性率（False discovery rate，FDR）校正，显著性水平 $P < 0.05$，导入质谱分析结果进行分析；CELLO V2.5 软件在 http：//cello.life.nctu.edu.tw/数据库中对全部蛋白进行亚细胞定位，从中筛选相关膜蛋白。

三、结果与分析

（一）蛋白相互作用网络图构建

Cytoscape 是目前蛋白质网络显示中广泛使用的一个可视化平台，通过集成 ClusterViz 蛋白质网络聚类分析和显示插件来实现 Cytoscape 新的算法和更多的网络分析功能。在该分析中，通过 Cytoscape V3.1.0（插件 ClusterViz V3.1.0）导入模块（111 个质谱分析结果），选择 Wikipathways、Reactome、GeneOntology、KEGG Pathway、基因功能富集性分析，经分析、计算结果显示：有 81 个符合要求的功能模块，并对该功能模块进行蛋白网络图谱的构建图 4-5 所示。由图中可以看出，A2M（α2 巨球蛋白）和 VIM（波形蛋白）与 CART 相互作用的相关系数最大。

（二）CART 相互作用蛋白的基因功能富集性分析

经 Cytoscape V3.1.0 导入 111 个蛋白组分对应的基因，进行基因功能富集性分析，结果如图 4-6 所示，共分为三大类 34 组：其中，分子功能占 11.73%，生物学过程占 46.93%，细胞组分占 41.34%。在卵泡发育过程中，CART 主要抑制卵泡 GCs 的增殖及 E_2 的分泌，从而发挥其负调控的功能，功能富集性分析结果显示，CART 与 NALP1、SERPINH1、PDIA6 的负调控应答（Negative regulation of response）功能提示，这些基因或蛋白可能对卵泡的发育产生重要影响。

（三）蛋白亚细胞定位分析

本研究中，LC-ESI-Q-TOF 质谱鉴定的 111 个蛋白质组分，经 CELLO V2.5 Subcellular Localization Prediction 数据库分析进行亚细胞定位，结果显示（图 4-7）：定位于细胞核的蛋白占 32.03%，细胞质蛋白占 21.09%，叶绿体蛋白 0.78%，胞外蛋白 21.09%，线粒体蛋白

14.84%，内质网蛋白占 2.34%，膜蛋白占 7.81%。其中，膜蛋白有 10个，分别为 A2M，C5，CNNM3，COX2，DDOST，HEATR5A，B3AT，

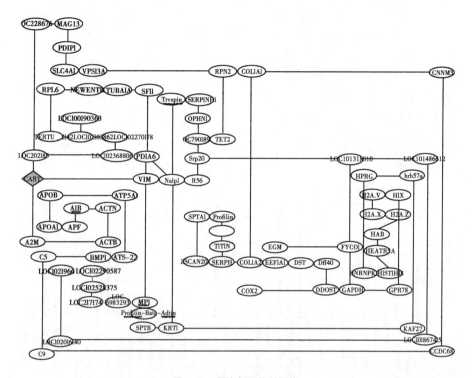

图 4-5 蛋白相互作用网络

Figure 4-5 Protein interaction network diagram

ADT2，RPN2，SLC4A1（表 4-4）。

表 4-4 膜蛋白分析

Table 4-4 Analysis of membrane proteins

Protein Accession	-10lgP	Coverage (%)	#Peptides	#Unique	Avg. Mass	Protein Name	Subcellular Localization Reliability	Protein Description
gi｜444518826	73.3	1	2	2	171 380	A2M	1.066	Alpha-2-macroglobulin
gi｜116608	29.07	0	1	1	188 877	C5	1.433	Complement C5
gi｜426226530	34.44	2	1	1	57 725	CNNM3	1.031	metal transporter CNNM3 partial
gi｜557467638	30.55	3	1	1	25 565	COX2	4.746	cytochrome c oxidase subunit II (mitochondrion)

（续表）

Protein Accession	−10lgP	Coverage (%)	#Peptides	#Unique	Avg. Mass	Protein Name	Subcellular Localization Reliability	Protein Description
gi | 545200782	47.93	2	1	1	70 759	DDOST	1.678	dolichyl-diphosphooligo-saccharide-protein glycosyltransferase subunit 2
gi | 351715661	30.07	0	1	1	235 762	HEATR 5A	2.243	HEAT repeat-containing protein 5A
gi | 30794360	228.8	19	15	15	104 375	B3AT	4.461	band 3 anion transport protein
gi | 351709909	80.96	9	3	1	32 879	ADT2	1.034	ADP/ATP translocase 2
gi | 296481161	35.03	2	1	1	69 214	RPN2	1.615	ribophorin II
gi | 296476229	228.8	19	15	15	104 375	SLC4A1	4.461	solute carrier family 4 anion exchanger member 1

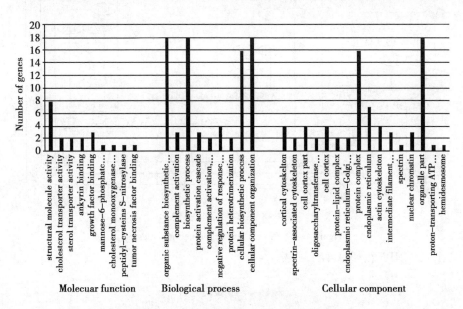

图 4–6 CART 相互作用蛋白 GO 分类

Figure 4–6 GO classification of protein interaction on CART

四、讨论

蛋白质组学与传统的蛋白质学科有本质的区别，是从生物体组织或其细胞的整体蛋白质入手进行研究，从生物有机体或细胞蛋白质整体活

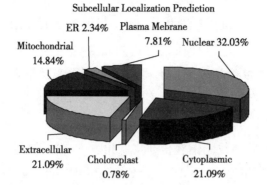

图 4-7 亚细胞定位预测

Figure 4-7 Subcellular localization prediction

动来揭示生命规律。与基因组学相比，蛋白质组学从更深入接近生命本质的层次去发现和探索生命活动规律及其重要生理病理现象的本质。PPI 作为细胞蛋白复合物和传导途径的基本要素，对蛋白功能的实现起着关键作用。Uetz 等（2000）利用酵母双杂交系统对酵母约 6 000 种 PPI 进行了大规模深入研究，并通过阵列筛选和文库筛选结果的对比，发现阵列筛选更为有效而文库筛选通量更大。Schwikowski 等（2000）对酵母的 2 709 种 PPI 进行了全面的分析，构建了两个包含 1 548 种蛋白以及 2 358 种蛋白的 PPI 网络图谱。Rain 等（2001）利用酵母双杂交系统筛选了 261 种幽门螺旋杆菌蛋白，明确了 1 200 种 PPI，构建了人类胃病病原体幽门螺旋杆菌的 PPI 图谱，将 46.6% 的蛋白组有机地联系起来。Walhout 等（2002）利用酵母双杂交系统研究了线虫生殖器的发育过程，通过酵母双杂交系统大规模地研究了 27 种已知的与线虫发育有关的蛋白，明确了 100 多个 PPI。

质谱法在真核生物蛋白检测和鉴定方面已广泛使用。尹晓（2001）通过 2-DE 从非肥胖和肥胖个体脂肪组织中获得 600 个蛋白图谱，经分析发现 96 个显著性差异蛋白点，并对其中 10 个差异蛋白点进行 MALDI-TOF-MS 质谱分析，获得的 7 个肽指纹图分别与 FGFR、lipin、IGFBP-3、aminopeptidase、蛋白酶 K 以及两种未知功能蛋白的酶

牛卵泡可卡因-
苯丙胺调节转录肽（CART）受体的筛选

解片段相匹配。陈述林（2004）选择人正常组（血液和心电图各项指标正常）、糖尿病组和肥胖组共12例（每组4例）动脉血管标本，通过SDS-PAGE垂直电泳和质谱分析发现，各组间差异表达蛋白点共124个，其中，109个蛋白表达上调，15个蛋白表达下调，t-Rorl蛋白在糖尿病组、肥胖组动脉血管组织的表达与对照组之间存在显著性差异，提示t-Rorl可能与血管病变的炎性反应有关。陈斌（2006）通过皮肤细胞蛋白的肽指纹图谱获得与皮肤角质形成相关的3个蛋白，分别为热休克蛋白HSP70、Tubulin α2和细胞骨架Actin gamma 1。解明然（2010）通过2-DE联用MALDI-TOF-MS对正常人及食管鳞癌患者（Ⅰ~Ⅲ期）血清蛋白质指纹图谱进行了研究，结果发现：过氧化物还原酶Ⅱ和TTR在患者血清中表达下调；结合球蛋白和血清淀粉样蛋白A在患者血清中表达上调；Apo-A1和细胞角蛋白1在局部晚期患者血清中表达上调。朱勇等（2012）通过液相色谱联用质谱分析技术对高生育力和低生育力人精子的差异蛋白进行了研究，结果共发现21个差异蛋白质点，其中，4个蛋白在低生育力人精子中高表达，其他皆在高生育力人精子中高表达。通过功能分析确认这些差异表达蛋白主要参与精子发生及受精、细胞的有丝分裂及凋亡和胚胎发育等过程。

从CART相互作用蛋白网络图谱可知，CART与A2M和VIM的相关系数最大。A2M基因位于牛第5号染色体，mRNA全长4 900bp，属于α-巨球蛋白家族，主要通过调节动物体内外源蛋白酶活性来参与多种生理功能。VIM（Vimentin）为波形蛋白，在间质组织广泛表达，对细胞完整性的维持及细胞应激反应的产生具有重要作用；研究也表明，在许多疾病尤其是一些恶性肿瘤发生过程中，均有VIM的异常表达，造成癌细胞的生长、迁移加速，预后效果较差。从功能上来看，对于CART与A2M和VIM的关系，还有待于进一步的深入研究。

通过对CART相互作用蛋白的功能富集性分析来看，CART与NALP1，SERPINH1，PDIA6具有负调控应答的功能，在动物卵泡发育的过程中，可能参与了GCs凋亡程序的启动，进一步促进了卵泡的闭锁。NALP1 属于 NOD 样受体家族成员，主要参与诱导细胞凋亡体的形成；在这个过程中，NALP1 与 caspse-2、caspase-3 和 caspase-9 相互作用，从而增强凋亡体的形成。F Liu 等（2004）在 HELA 细胞体外培养研究中证实，NALP1 通过诱导激活 caspase-3 促进了 HELA 细胞的凋亡；CF Lo 等（2008）研究发现瞬时表达的重组 NALP1 神经元诱导 caspase-3 活化和细胞凋亡，表明 NALP1 在受损神经元内可能参与了细胞凋亡蛋白酶活化的调节。SERPINH1（Serine peptidase inhibitor H1）为丝氨酸蛋白酶抑制剂家族成员之一。目前，对该蛋白功能的研究主要集中在两个方面，与成骨不全症的形成和肿瘤细胞生长的关系。HE Christiansen 等研究发现，在人和小鼠体内 SERPINH1 基因发生变异，可导致蛋白装配、折叠和分泌过程异常，导致胎儿发育受阻骨骼形成异常；YY Yu 等 siRNA 干扰技术研究发现，SERPINH1 基因沉默后对模型小鼠的肿瘤生长有促进作用，即 SERPINH1 具有肿瘤抑制作用；PDIA6 分布于内质网、高尔基体中心区域，该蛋白参与二硫键形成、异构化及氧化还原过程，并在蛋白质转录后修饰、装配、折叠及运输中发挥作用。目前，PDIA6 与细胞生长调节的相关性主要体现在肿瘤研究方面，PDIA6 通过信号通路激活及蛋白折叠反应激活，促进肿瘤的转移和侵袭。JA Vekich 等（2012）通过 ATF6 转基因小鼠模型研究表明，PDIA6 通过促进二硫键的形成，增强蛋白质折叠，可对缺血性 ATF6 小鼠的心脏起保护作用。

为了确定 CART 的受体，前人已做了大量研究，到目前基本明确了 CART 受体的相关特性，但受体蛋白仍未确定。本研究通过 Co-IP 技术对 CART 相互作用蛋白进行了研究，为进一步确定 CART 受体奠定了基础。

小结

1. 利用兔抗 CART 一抗沉淀 CART 及其相互作用蛋白，经 Protein

G - Agarose 磁珠吸附分离，获得符合质谱分析的蛋白复合体。

2. 经 LC - ESI - Q - TOF 质谱分析及数据库比对，共鉴定蛋白质组分 111 个。

3. 经 Cytoscape V3.1.0 分析，有 81 个蛋白符合功能模块并构建蛋白相互作用网络图。

4. 经 CELLO V2.5 Subcellular Localization Prediction 数据库分析，定位于细胞核的蛋白占 32.03%，细胞质蛋白占 21.09%，叶绿体蛋白 0.78%，胞外蛋白 21.09%，线粒体蛋白 14.84%，内质网蛋白占 2.34%，胞膜蛋白占 7.81%，其中，膜蛋白有 10 个。

第五章

G 蛋白偶联受体的筛选及同源建模

第一节 G 蛋白偶联受体的预测

一、概述

神经肽是生物体内的一类生物活性多肽，大多分布于神经组织，也可存在于其他组织，按其分布不同分别起递质（Transmitter）、调质（Modulator）或激素（Hormone）的作用。神经肽按发现部位分类，可分为：垂体肽、下丘脑神经肽、脑肠肽、内源性阿片肽和其他肽类。CART 是动物下丘脑分泌的神经肽，属于经典激素类。随着 DNA 重组技术的广泛应用，大多数神经肽的受体已克隆成功。目前已发现的神经肽的受体，除心房钠尿肽（ANP）受体外，所有已克隆的神经肽受体都属于鸟核苷酸调节蛋白，即 G 蛋白偶联的受体。与非肽类 G 蛋白偶联的受体一样，这类受体的共同特征是受体蛋白的肽链形成 7 个 α 螺旋区段并跨膜 7 次，在大多数情况下要通过细胞内的第二信使产生效应。ANP 受体比较特殊，其本身就是膜上的鸟苷酸环化酶（cGMP），该受体激活时直接引起细胞内 cGMP 含量增加，不需要 G 蛋白的介导。

同时，科研人员通过 CART 信号传导途径和细胞结合试验研究，虽然 CART 受体的分子特性尚未明确，但 CART 的生物学作用是由受体介导的。CART - GFP 融合蛋白与 Hep G2 细胞和分离的下丘脑细胞的结合

牛卵泡可卡因－苯丙胺调节转录肽（CART）受体的筛选

已有报道；CART 与 AtT20 细胞、PC12 细胞和原代培养大鼠伏隔核细胞的特定饱和结合也已证明，当 GTP 存在时 CART 结合会减少，但不受 ATP 类似物存在的影响；此外，在研究 CART 在颗粒细胞、AtT20 细胞和海马神经元的作用时，当用和 CART 受体信号有关的 o/i 子类抑制性 G 蛋白的抑制剂预处理细胞时，细胞上的 Go/i 信号终止。所有的研究证据表明 CART 的生物学作用可能是通过 G 蛋白偶联受体介导的。Joseph 等（2013）在研究牛卵泡颗粒细胞 E_2 产生过程中 CART 信号通路时，也证明 CART 生物学作用的实现是由其受体介导的，其受体可能是 G 蛋白偶联受体。而 G 蛋白偶联受体是一类膜蛋白受体的统称。

因此，在研究过程中，通过筛选 G 蛋白偶联受体来缩小 CART 受体的筛选范围。

二、试验材料

（一）供试材料

高通量测序结果 ODF1/ODF2 >2 的 1 331 个差异表达基因中筛选出的 188 个膜蛋白，以及 Co – IP 和质谱分析中（111 个蛋白）筛选出的 10 个膜蛋白，共 195 个（3 个重复）膜蛋白。

（二）生物信息数据库和软件

NCBI（蛋白质数据库 http：//www.ncbi.nlm.nih.gov/protein/）。

HMMTOP V2.0（Prediction of transmembrane helices and topology of proteins Version 2.0, http：//www.enzim.hu/hmmtop/html/submit.html）。

三、试验方法与结果

本研究中，应用 HMMTOP V2.0（Prediction of transmembrane helices and topology of proteins Version 2.0），对深度测序结果中 ODF1/ODF2 >2

第五章
G 蛋白偶联受体的筛选及同源建模

的差异表达基因中筛选出的 188 个膜蛋白，以及 Co‑IP 和质谱分析中（111 个蛋白）筛选出的 10 个膜蛋白，共 195 个（3 个重复）膜蛋白中筛选 G 蛋白偶联受体，结果见图 5‑1 所示，共筛选出含有 7 个跨膜 α 螺旋区的 38 个膜蛋白，因此，这 38 个膜蛋白属于 G 蛋白偶联受体（表 5‑1）。

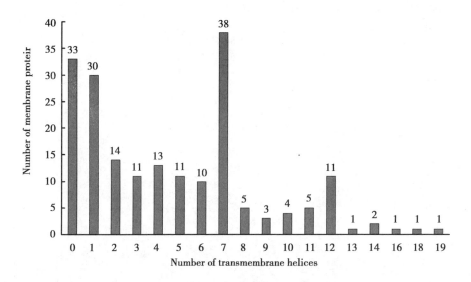

图 5‑1 蛋白跨膜 α 螺旋的数量分布

Figure 5‑1 Proteins distribution of transmembrane alpha helix

表 5‑1 38 个 G 蛋白偶联受体特征

Table 5‑1 Feature of G protein coupled receptors

Protein Name	Length	Transmembrane Helices	Protein Description
CMKLR1	362	36‑60, 73‑96, 111‑130, 151‑172, 220‑239, 256‑276, 291‑315	chemokine‑like receptor 1
GPR156	808	48‑72, 85‑107, 120‑142, 167‑187, 225‑247, 260‑280, 289‑311	G‑protein coupled receptor 156
APLNR	400	29‑53, 66‑87, 104‑123, 144‑167, 204‑228, 249‑273, 290‑314	apelin receptor

牛卵泡可卡因-苯丙胺调节转录肽（CART）受体的筛选

（续表）

Protein Name	Length	Transmembrane Helices	Protein Description
MRGPRF	343	45-69, 78-102, 117-139, 160-184, 199-222, 241-265, 278-295	mas-related G-protein coupled receptor member F
GPR4	362	23-42, 53-72, 93-113, 134-154, 181-200, 225-244, 271-289	G-protein coupled receptor 4
MBOAT4	435	18-37, 50-74, 162-181, 212-231, 345-369, 380-397, 404-428	membrane bound O-acyltransferase domain containing 4-like
LPAR6	341	18-42, 53-72, 93-112, 135-154, 181-205, 228-252, 273-292	lysophosphatidic acid receptor 6
GPR116	1343	626-649, 668-690, 703-727, 746-766, 787-811, 836-856, 865-884	G-protein coupled receptor 116
CCBP2	384	49-73, 86-109, 118-142, 163-187, 218-242, 255-274, 301-318	chemokine-binding protein 2
LGR5	907	562-581, 594-616, 639-660, 685-704, 725-744, 769-791, 804-823	leucine-rich repeat-containing G-protein coupled receptor 5
GPR85	370	24-46, 57-81, 96-118, 135-154, 185-209, 286-305, 320-342	G-protein coupled receptor 85
GPR39	454	25-47, 72-89, 114-131, 154-176, 224-246, 281-298, 321-343	G-protein coupled receptor 39
CHRM5	532	31-55, 68-92, 107-126, 147-166, 193-212, 443-464, 479-498	muscarinic acetylcholine receptor M5
PQLC3	201	4-26, 35-52, 61-83, 92-108, 115-132, 141-161, 170-191	PQ-loop repeat-containing protein 3
LOC782475	323	26-50, 59-78, 95-119, 144-166, 197-216, 241-260, 273-292	olfactory receptor 2W6
GPR124	957	607-626, 633-651, 660-678, 710-728, 749-767, 844-862, 871-889	G-protein coupled receptor 124
GPER	354	38-62, 75-93, 106-130, 155-175, 198-217, 240-260, 287-306	G-protein coupled estrogen receptor 1
AGTR2	362	45-68, 77-101, 118-139, 160-179, 210-233, 258-282, 297-321	type-2 angiotensin II receptor
P2RY6	328	27-49, 62-83, 104-122, 143-167, 194-216, 241-265, 286-304	P2Y purinoceptor 6
FZD4	537	221-243, 256-275, 302-325, 346-365, 392-414, 433-452, 479-498	frizzled-4 precursor
SRD5A3	318	17-41, 79-102, 125-141, 160-178, 199-215, 251-270, 275-294	steroid 5 alpha-reductase 3-like
GPR137C	421	31-55, 76-98, 117-139, 156-180, 193-216, 239-262, 293-314	integral membrane protein GPR137C

(续表)

Protein Name	Length	Transmembrane Helices	Protein Description
LOC526047	316	31–55, 64–83, 96–115, 128–152, 207–231, 242–265, 278–297	olfactory receptor 12
LOC514434	315	25–49, 66–85, 100–119, 138–162, 197–221, 240–259, 272–291	olfactory receptor 5V1
EMR1	900	616–640, 649–668, 677–699, 724–743, 764–788, 809–829, 838–862	EGF–like module–containing mucin–like hormone receptor–like 1
LOC782554	307	26–48, 59–78, 97–119, 144–166, 197–220, 241–260, 273–292	olfactory receptor 2W6
LOC524771	295	8–30, 53–70, 103–122, 133–150, 163–180, 193–210, 237–254	uncharacterized protein
S1PR4	385	45–69, 82–106, 125–142, 167–191, 206–228, 253–274, 293–310	sphingosine 1–phosphate receptor 4
TEDDM1	509	8–26, 57–75, 100–118, 129–146, 159–177, 188–206, 233–250	transmembrane epididymal protein 1
LOC787041	310	26–48, 61–83, 96–119, 132–154, 192–216, 237–260, 273–292	olfactory receptor 18
GPR142	480	178–199, 212–231, 252–271, 296–315, 336–355, 376–395, 416–435	G–protein coupled receptor 142
LOC512627	321	28–52, 65–89, 100–124, 147–166, 205–229, 242–266, 277–296	olfactory receptor 2AT4
LOC614050	312	26–49, 62–86, 101–119, 140–164, 200–224, 237–261, 274–292	olfactory receptor 2D3
RRH	337	27–51, 62–85, 100–121, 142–165, 186–210, 241–265, 274–297	visual pigment–like receptor peropsin
CASR	1085	613–636, 651–671, 684–707, 722–744, 771–793, 808–827, 840–863	extracellular calcium–sensing receptor precursor
ADRA1B	577	48–70, 83–105, 120–141, 162–183, 202–221, 295–314, 329–348	alpha–1B adrenergic receptor
FZD3	666	205–226, 239–258, 285–307, 328–347, 374–396, 421–440, 477–500	frizzled homolog 3
APH1B	257	4–23, 32–54, 71–90, 115–134, 155–178, 187–206, 213–235	gamma–secretase subunit APH–1B

四、讨论

G 蛋白偶联受体（G protein–coupled receptors，GPCRs）是最大的一类细胞膜表面受体。当 G 蛋白偶联受体被胞外信号分子刺激后可以

61

牛卵泡可卡因－
苯丙胺调节转录肽（CART）受体的筛选

激活细胞内的信号转导通路，并最终诱导细胞反应（例如：神经传递、生长、发育、细胞分化、炎症、免疫反应等）。因此，这类受体与多种疾病的形成、发生息息相关，从而成为了新药开发和研究的重要靶点。

在 Brian K 和 Raymond S 等人的努力下，现在已经比较清楚地了解了 G 蛋白偶联受体的三维拓扑结构。除了 7 次跨膜螺旋，受体分子还包含 3 个胞内环以及 3 个胞外环结构。在 3 个胞内环中，一般前两个胞内环较短，没有特定的生理功能；第 3 个胞内环则变化较大，在不同的受体中长度可能从几个氨基酸到一百余个氨基酸残基不等。3 个胞内环中，通常第 3 个胞内环的功能最为重要，涉及 G 蛋白偶联受体与 G 蛋白的识别及相互作用；但同时也由于其柔性太大，通常会被可溶蛋白片段取代或者缺乏均一的构象，所以到目前为止具体的结构仍不清楚。与此类似的 3 个胞外环中，同样第一个以及第 3 个胞外环较短，没有明确的生物学功能；而第 2 个胞外环在不同的受体中差异比较大，有可能形成 α 螺旋，也有可能是 β 折叠，还有一部分受体的环区较短，仅形成简单的无规则环区。第 2 个胞外环通常会形成一个类似盖子的结构罩住底物结合位点，不但有可能对底物的进出造成影响，其部分残基也参与了与底物的结合，它与螺旋一起形成了特异的选择机制。

生物信息学是一门数学、统计、计算机与生物医学交叉结合的新兴学科，其利用计算机技术研究生物系统规律，也是分子生物学与信息技术的结合体。生物信息学的研究材料和结果就是各种各样的生物学数据，其研究工具是计算机，研究方法包括对生物学数据的搜索（收集和筛选）、处理（编辑、整理、管理和显示）及利用（计算、模拟）。在预测膜蛋白中的 G 蛋白偶联受体时，就是通过生物信息学技术预测蛋白序列中是否存在 7 次跨膜螺旋来确定 G 蛋白偶联受体。

第二节 CART 及 G 蛋白偶联受体的同源建模

一、概述

同源建模法是基于蛋白质晶体结构和蛋白质同源性的空间结构预测方法。通过对 PDB 数据库中的蛋白质进行立体结构比对的定量研究获取，任何蛋白质只要其氨基酸序列达到一定长度并具有同源性超过 30% 的已知晶体结构的蛋白质结构域存在，就可以获得可信合理的立体三维结构。因此，对于未知晶体结构的蛋白质，可设法找出已知晶体结构并与其高度同源的蛋白质，从而建立该蛋白的三维立体结构模型。本试验通过缩小 CART 受体的选择范围，应用 SWISS – MODEL 和 MODELLER 对配体、受体分子进行同源建模，为蛋白质间的分子对接提供基础。

二、试验材料

（一）供试材料

CART 和 38 个 G 蛋白偶联受体的氨基酸序列均来源于美国国家生物技术信息中心（National Center of Biotechnology Information，NCBI）数据库，序列均为 FASTA 格式。下面以 CART 和 CHRM5 的 FASTA 格式为例展示，其他蛋白 FASTA 格式在此省略。

＞gi｜56119134｜ref｜NP_ 001007821.1｜ cocaine – and amphetamine – regulated transcript protein precursor ［Bos taurus］

　　MESPRLRLLPLLGAALLLLLPLLGALAQEDAELQPRALDIYSAVEDASHEKELIEALQEVLKKLKSRIPIYEKKYGQVPMCDAGEQCAVRKGARIGKLCDCPRGTSCNSFLLKCL

＞gi｜555960885｜ref｜XP_ 005892543.1｜ PREDICTED：muscarinic acetylcholine receptor M5 ［Bos mutus］

牛卵泡可卡因－苯丙胺调节转录肽（CART）受体的筛选

MEGEPYHNASTVNSTPVSHQPLERHGLWEVITIAAVTAVVSLITIVGN
VFVMISFKVNSQLKTVNNYYLLSLACADLIIGIFSMNLYTTYILMGRWALGS
LACDLWLALDYVASNASVMNLLVISFDRYFSITRPLTYRAKRTPKRAGIMIG
LAWLISFILWAPAILCWQYLVGERTVPPDECQIQFLSEPTITFGTAIAAFYIPV
SVMTILYCRIYRETEKRTKDLADLQGSDSVAEAEKRKSAHQALLRSCFSCP
WPTLAQRERNQASWSSSRRSTSITGKPSQATGPSTEWAKAEQLTTCSSYPSSE
DEDKPTTDPVFQVVYKSQAKESPREELSAEETKQTFVNAQSEKNDYDTPKY
FLSPASAHGPKSQKWVAYKFRLVVKAEGTHDTNNGCRKVKIMPCSFPVSK
DPSTKGLDPNLSHQMTKRKRMVLVKERKAAQTLSAILLAFIITWTPYNIMV
LVSTFCDKCVPVTLWHLGYWLCYVNSTVNPICYALCNRTFRKTFKMLLFC
RWKKKKVEEKLYWQGNSKLP

（二）生物信息数据库和软件

NCBI（蛋白质数据库 http：//www.ncbi.nlm.nih.gov/protein/）。

RCSB PDB（PDB 蛋白质数据库 http：//www.rcsb.org/pdb/results/static.do）。

SWISS – MODEL（Automated protein structure homology – modelling server，http：//swissmodel.expasy.org/interactive）。

MODELLER 蛋白质序列比对软件。

三、试验方法与结果

（一）SWISS – MODEL 同源建模

研究人员通过对蛋白质空间结构的大量比较发现，蛋白质三级结构比其相应一级结构序列的保守性更高。氨基酸残基序列同源性达到 50% 的蛋白质，活性中心约有 90% 的 Cα 原子偏差小于 2.5Å，均方根偏差小于 1Å。由于氨基酸残基发生替换通常出现在蛋白质表面的回折区域，蛋白质主链结构，尤其是疏水结合中心的结构，受氨基酸残基序

第五章 G 蛋白偶联受体的筛选及同源建模

列变化的影响非常小。因此，用蛋白质数据库来预测目标蛋白的立体空间结构是可靠的。SWISS – MODEL 和 PDB 数据库同源建模通常包括以下步骤。

（1）通过 SWISS – MODEL 运行 PDB 数据库，搜索序列同源性（注意：非相似性）最高的且超过 30% 的蛋白质。

（2）由氨基酸序列同源性较高的已确认三级结构的蛋白质信息确定目标蛋白质的保守区结构（Structural conservative regions，SCR）。

（3）通过序列对比，从序列同源蛋白结构的 SCR 确定目标蛋白序列的 SCR。

（4）在目标蛋白序列的 SCR 上构建主链。

（5）用氨基酸序列同源蛋白的结构或构象搜索方法构建目标蛋白的结构可变区（Loop）。

（6）通过（4）和（5）获得的目标蛋白质 Cα 原子坐标，构建分子三维坐标并完成侧链组装。

（7）结构优化和评估。

（8）输出 PDBQT 文件用于后续分析。

当目标蛋白与参考蛋白氨基酸序列同源性大于 30% 时，可在原子水平的分辨精度上成功预测目标蛋白的三维空间结构。应用 PDB 蛋白质数据库进行同源建模，大约有 30% 的氨基酸序列与一个或多个已知晶体立体结构的序列有较高（大于或等于 30%）的同源性，部分蛋白完全可以通过 PDB 数据库同源建模构建其三维立体空间结构。

通过 SWISS – MODEL 运行 PDB 数据库进行同源建模时，配体 CART 和 12 个 G 蛋白偶联受体可以通过 PDB 数据库同源建模（表5 – 2）。

表 5 – 2　13 个 SWISS – MODEL 建模蛋白

Table 5 – 2　13 proteins MODEL is established through the SWISS – MODEL

Protein Name	Template/PDBID	Seq Identity
CART	1hy9.1.A	100%

(续表)

Protein Name	Template/PDBID	Seq Identity
CMKLR1	4n6h.1.A	31.43%
APLNR	3odu.1.B	31.25%
LPAR6	3vw7.1.A	31.93%
CHRM5	4mqs.1.A	59.22%
AGTR2	4n6h.1.A	32.04%
S1PR4	3v2w.1.A	42.91%
RRH	3ayn.1.A	30.07%
CASR	2e4x.1.A	33.65%
ADRA1B	3d4s.1.A	34.98%
LGR5	4kng.1.A	93.12%
GPR39	4buo.2.A	32.79%
CCBP2	4mbs.1.A	35.14%

下面以配体 CART 为例介绍 PDB 结构数据库同源建模过程。

1. 目标序列的搜索建模

利用 SWISS – MODEL 中的 Blast Search 模块，将牛 CART（116 AA）蛋白质序列以 FASTA 格式输入，搜索 PDB 数据库中可同源建模的蛋白空间立体结构。搜索结果如表 5 – 3 所示，可见，牛 CART 立体结构与人源 1hy9.1.A 同源性最高，为 100%，可作为蛋白模型建立分子坐标。

图 5 – 2（a）为 1hy9.1.A 的立体结构，其中，包含有 2 个 β 折叠及 6 个半胱氨酸形成的 3 个二硫键，图 5 – 2（b）为 1hy9.1.A 的分子立体模型。

2. 模型的评价与分析

通过 QMEAN4 评分函数对目标蛋白总体模型质量和每个氨基酸残基的模型质量（Global and per – residue model quality，GMQE）进行评价，为了提高评分性能，SWISS – MODEL 通常对 QMEAN 个别参数的权重进行了调整。

QMEAN4 评分函数对目标蛋白总体模型质量评价如图 5 – 3 所示，

表5-3 CART同源蛋白搜索
Table 5-3 Search results of CART homologous proteins

Template	Seq Identity	Oligo-state	Found by	Method	Resolution	Seq Similarity	Coverage	Description
1hy9.1.A	100.00	monomer	HHblits	NMR	NA	0.64	0.35	COCAINE AND AMPHETAMINE REGULATED TRANSCRIPT PROTEIN
1dip.1.B	20.41	homo-dimer	HHblits	NMR	NA	0.30	0.42	DELTA-SLEEP-INDUCING PEPTIDE IMMUNOREACTIVE PEPTIDE
1dip.1.B	20.41	homo-dimer	HHblits	NMR	NA	0.30	0.42	DELTA-SLEEP-INDUCING PEPTIDE IMMUNOREACTIVE PEPTIDE
3bii.1.B	23.33	hetero-oligomer	HHblits	X-ray	2.50Å	0.36	0.26	Molybdopterin-converting factor subunit 2
2q5w.1.A	23.33	hetero-oligomer	HHblits	X-ray	2.00Å	0.35	0.26	Molybdopterin-converting factor subunit 2
2qie.1.A	23.33	hetero-oligomer	HHblits	X-ray	2.50Å	0.35	0.26	Molybdopterin-converting factor subunit 2
2qie.1.C	23.33	hetero-oligomer	HHblits	X-ray	2.50Å	0.35	0.26	Molybdopterin-converting factor subunit 2
2wp4.1.A	31.03	homo-dimer	HHblits	X-ray	2.49Å	0.37	0.25	MOLYBOOPTERIN-CONVERTING FACTOR SUBUNIT 2 1
2wp4.1.B	31.03	homo-dimer	HHblits	X-ray	2.49Å	0.37	0.25	MOLYBOOPTERIN-CONVERTING FACTOR SUBUNIT 2 1
1nvj.1.A	24.14	homo-dimer	HHblits	X-ray	2.15Å	0.37	0.25	Molybdopterin converting factor subunit 2
1nvj.1.B	24.14	homo-dimer	HHblits	X-ray	2.15Å	0.37	0.25	Molybdopterin converting factor subunit 2
1nvj.3.A	24.14	homo-dimer	HHblits	X-ray	2.15Å	0.37	0.25	Molybdopterin converting factor subunit 2
3rpf.1.A	17.24	hetero-oligomer	HHblits	X-ray	1.90Å	0.33	0.25	Molybdopterin synthase catalytic subunit

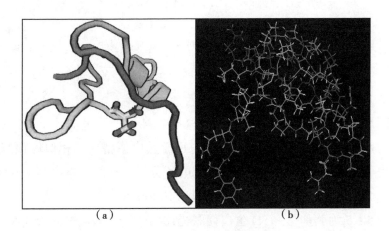

图 5-2 CART 立体结构

Figure 5-2 CART three-dimensional structure

横坐标为蛋白大小（埃米），纵坐标为标准化的 QMEAN4 得分，总体模型是与 PDB 非冗余结构集合的对照。由图可以看出，总体模型中绝大

> 牛卵泡可卡因-
> 苯丙胺调节转录肽（CART）受体的筛选

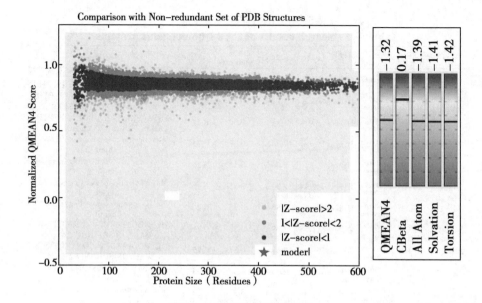

图 5-3 总体模型质量评价
Figure 5-3 Global model quality estimate

部分原子 Z-score 的绝对值小于 1，只有少数原子 Z-score 的绝对值大于 2；整体 QMEAN4 平均得分的绝对值为 1.32，全部原子 Z-score 平均得分的绝对值为 1.39；在溶解状态下，全部原子 Z-score 平均得分的绝对值为 1.41；在侧链发生旋转时，全部原子 Z-score 平均得分的绝对值为 1.42。

图 5-4 是对构建模型中每个氨基酸残基模型质量的评价，图中横坐标为残基数（76～116），纵坐标为目标蛋白模型局部相似性预测值。从图中可以看出，每一个氨基酸残基的评分均为正值，这说明构建模型中每一个氨基酸残基都处于可信合理的空间位置。

图 5-5 为其他 12 个 G 蛋白偶联受体构建模型的质量评价曲线，从图中可以看出，所构建模型的每一个氨基酸残基的评分均为正值，这说明构建模型中每一个氨基酸残基都处于可信合理的空间位置。

经过模型的构建和评价，将该模型以 PDBQT（分子坐标）或 PDBID 文件输出，保存备用。

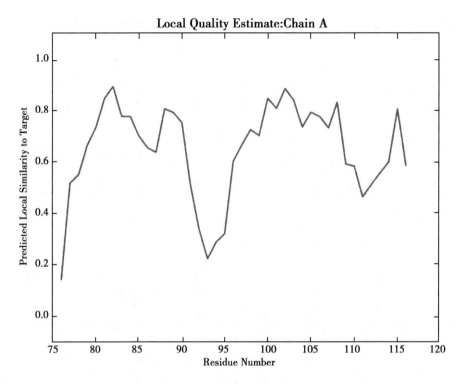

图5-4 CART模型质量评价

Figure 5-4 CART model quality estimate

(二) MODELLER 同源建模

通过 SWISS-MODEL 和 PDB 数据库可以对配体 CART 和 12 个 G 蛋白偶联受体同源建模,但不能够对其他 26 个 G 蛋白偶联受体直接建模,需通过加利福尼亚大学开发的同源建模软件 MODELLER 9.13 结合 NCBI 蛋白质 BLAST 和 FASTA 搜索引擎为其他 26 个 G 蛋白偶联受体建立立体空间模型,并对该模型进行评价。

下面以 GPR116 为例介绍同源建模软件 MODELLER 建模过程。

1. GPR116 的 MODELLER 建模方法

(1) 将目标蛋白 GPR116 的氨基酸序列转换成 MODELLER 可识别的如下文件格式。

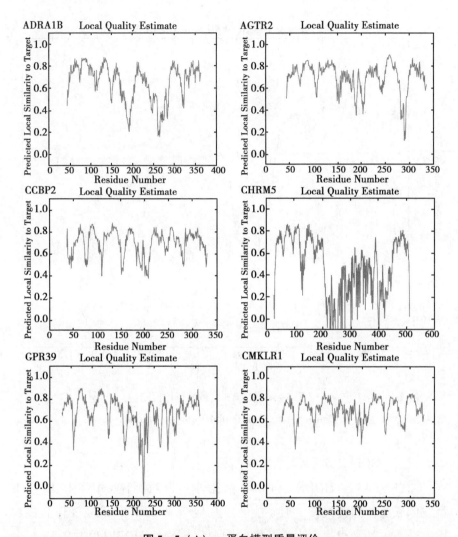

图 5-5（A） 蛋白模型质量评价

Figure 5-5（A） Protein model quality estimate

第二行必须把第一个字段 sequence 写出，需要注意的是这一行的 10 个字段中有 9 个分号；第三行开始为 GPR116 的氨基酸序列，氨基酸序列结尾用"*"代表序列结束。

（2）将 BLAST 搜索出来的 PDB 库中同源性最高的 4 条以上序列准

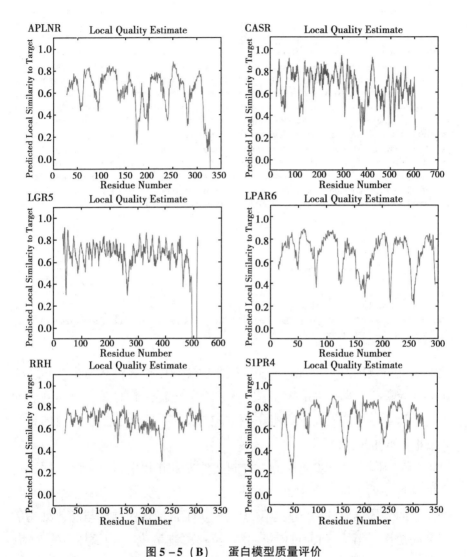

图 5-5（B） 蛋白模型质量评价

Figure 5-5（B） Protein model quality estimate

备好，以下是将 PDB 数据库中的氨基酸序列转化为二进制的文件格式，可以加快搜索同源序列的速度。全部过程使用 Python2.7 编程语言，转换的运行脚本如下。

第四行中 sequence_ db（） 是新建一个存储大量氨基酸序列的库对

牛卵泡可卡因－苯丙胺调节转录肽（CART）受体的筛选

```
>P1 ; TvLDH
sequence : : : : : : : : : :0.00: 0.00
MKSPKRTTLCFVIIAIYSSQATLSLNLESIVYHSLLHEHELMEEDPVRQKRAVAISSPAAQEFTVDVEISFENASFLEPIKAYLNNLSFPIQGNDTDLATDILSIEVSTVC
RPTGNEIWCSCETGYEWPQESCLHNLTCQCHDGSSAGHHCSCLKGLPPKGPFCQLQGDIILRMSVRLNVGFQEDLKNSSSALYKSYKTDLETAFWKGYSILPGFKLVTVTG
FRPGSVVVTYEITTTTTASPALMRKANEKVTQNLNQTYKMESSSYQVTYSNETKFSITPEIIFEGDTVTLMCENEVFSTNVSWTQGETGLDIQNSTEVSIYTSVLNNMTSV
SRLIIHNITRGFMGEYICKLTLDIFEYASKRKVEVIPIQILASEATEVMCDNNPVSLHCCSEVNVNWSQIEWKQEGSISVPGNAEIDTDSGCSRYTLVANGTQCPSGSSGT
TVVTTCEFISAHGARGRKDIEVTFMSVAKLNITPDLISVSEGQSFFIKCISDMSNYDEMYWNTSAGTKIYKRFYTTRRYPDRAESVLTVKTSTREWNGTYHCVFRYKNLYR
VMAKDVIVYPLPLESNTMVDPVEAAIPCHSSHHIRCCIEEDSDYRVIFQVGSSSFPAVKEVNEKQVCYKHHFVADSVSCPENVDVFCHFANAANNSVQSPSMKLNLVPGES
ITCQDPILGVGEPGKVIQKLCRFSNITRSRESPIGGIITYRCVGSQWQIERNDCISAPINALLQLAKALVKSPTQDEKLPSYLGNLSISTKQVKLEVDSFSGTLGAHINIL
DKFSTVPTQVNSEMMTHVLSTVHIILGKSALNTWKMSQQQKTNYSSQLLDSVERFSRVLRAEDSTISQPNVQMKSMVIKPGHPQSYRQSFIFLDSDLWGNVTINGCQLEHL
QPDSFVVTVAFPTLKTILDRDVPGQNFANSLVMTTTVSRSITTPFRILMTFKNNHPLGGIPQCVFWNFDLANHTGGWDSSGCYVKEVTEDSVSCSCEEHLTSFSILMSPDSP
DPDSLLKILLDIISYIGLCFSILSLAACLVVEAVVWKSVTKNRTSSMRHICIVNITASLLVADVWFIVAAAATHDHRFPLNETACVAATFFVHFFYLSVFFWMLTLGLMLFY
RLVFILHDTSKSIQKTIAFSLGYGCPLVISIITVGATQPREVYMRKNACWLNWEDTKALLAFVIPALIIVVVNMTITVVVITKILRPSVGDKPSKQEKSSLFQVAKSIGVL
TPLLGLTWGFGLATVFQGSNAVFHIVFTLLNAFQGLFILLFGCLWDQKVQEALLNKYSRSKWSSQHSKSTSIGSSTPVFSMSSPISRRFNNLFGKTGTYNVSTPETTSSSL
ENTSSAYSLLN*
```

```
from modeller import *
log.verbose()
env = environ()
sdb = sequence_db(env)
sdb.convert(seq_database_file='pdb_95.pir',seq_database_format='PIR',chains_list='ALL',minmax_db_seq_len=(30,
40000),cleansequences=True,outfile='pdb_95'.hdf5')
```

象；第五行调用对象 convert（）方式，以 PIR 格式读入 pdb_ 95. pir, 用 minmx_ db_ seq_ len 参数筛选并去除序列数少于 30 和大于 4 000 的序列，以 clean_ sequences 剔除非标准残基，文件最后以二进制形式写入 pdb_ 95. hdf5 文件中。

（3）用第（2）步生成的二进制文件搜索并找出和目标序列同源性高的结构序列。该文件中，第 2、11、12 列最为关键，第 2 列为 PDB 数据库中的晶体结构的 ID 号，第 11 列设定了目标蛋白结构与晶体结构的序列同源性，第 12 列为目标蛋白结构与数据库晶体结构叠加后的 E 值。

（4）从第（3）步的输出文件中找出和目标蛋白同源性较高的模版，评价模版结构和序列同源性。

下图为 log 文件输出结果，其中，上部分为序列同源性的比较：对角线内容为不同结构的残基数，上三角为相同残基数，下三角二者序列的同源性；下部分为结晶的分辨率。由于 1y7t 的结晶分辨率为 1.6Å，同源性为 61%，故选择 1y7t 作为立体结构模版。

第五章
G 蛋白偶联受体的筛选及同源建模

```
from modeller import *
log.verbose( )
env = environ( )
# Read in the sequence database in binary format
sdb = sequence_db(env,seq_database_file='pdb_95.hdf5',seq_database_format='BINARY',chains_list='ALL')
# Read in the target sequence in PIR alignment format
aln = alignment(env)
aln.append(file='TvLDH.ali',alignment_format='PIR',align_codes='ALL')
# Convert the input sequence "alignment"into profile format
prf = aln.to_profile( )
# Scan sequence database to pick up homologous sequences
prf.build(sdb,matrix_offset=-450,rr_file=' $ {LIB}/blosum62.sim.mat',gap_penalties_1d=(-500,-50),
    n_prof_iterations=1,check_profile=False,max_aln_evalue=0.01)
# Write out the profile in text format
prf.write(file='build_profile.prf',profile_format='TEXT')
# Convert the profile back to alignment format
aln = prf.to_alignment( )
#– Write out a PIR alignment file
aln.write(file='build_profile.ali',alignment_format='PIR')
```

```
from modeller import *
env = environ( )
env.io.atom_files_directory=['.','../atom_files']
# Make a simple 1:1 alignment of 7template structures
aln = alignment(env)for(pdb,chain)in(('1b8p','A'),('1y7t','A'),('1civ','A'),('5mdh','A'),('7mdh','A'),('3d5t','A'),('1smk','A')):
    m = model(env,file=pdb,model_segment=('FIRST:'+chain,'LAST:'+chain))
    aln.append_model(m,atom_files=pdb,align_codes=pdb+chain)
# Sequence alignment
aln.malign( )
# Structure alignment
aln.malign3d( )
# Report details of the sequence/structure alignment
aln.compare_structures( )
aln.id_table(matrix_file='family.mat')
env.dendrogram(matrix_file='family.mat',cluster_cut=-1.0)
```

（5）目标结构与模版结构进行 align，为 MODELLER 构建目标结构设定参数；MODELLER 在设定参数时会具体到模版结构信息，因此，最好用 align2d（ ）动态算法来进行序列与序列的 align。

（6）创建比较模型。通过 MODELLER 中 automodel，将 alignment

73

牛卵泡可卡因－苯丙胺调节转录肽（CART）受体的筛选

Sequence identity comparison (ID_TABLE):

   Diagonal   ... number of residues;
   Upper triangle   ... number of identical resdues;
   Lower triangle   ... & sequence identity,id/min(length).

	1b8pA @1	1y7tA @1	1civA @2	5mdhA @2	7mdhA @2	3d5tA @2	1smkA @2
1b8pA @1	327	201	146	152	152	249	50
1y7tA @1	61	327	158	170	160	210	58
1civA @2	46	48	374	140	304	148	55
5mdhA @2	46	52	42	333	140	164	58
7mdhA @2	46	49	87	42	351	148	50
3d5tA @2	78	65	46	51	46	321	50
1smkA @2	16	19	18	19	16	16	313

Weighted pair-qroup average clustering based on a distance matrix:

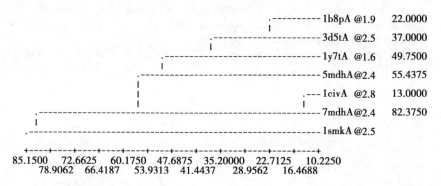

```
from modeller import *

env = environ( )

env.io.atom_files_directory=['.','../atom_files']

aln = alignment(env)

# Read in the 1y7t template structure and add to alignment

mdl = model(env,file='1y7t',model_segment=('FIRST A','LAST:A'))

aln.append_model(mdl,align_codes='1y7tA',atom_files='1y7t')

# Add in the TvLDH sequence

aln.append(file='../stepl_search/TvLDH.ali',align_codes='TvLDH')

# Sequence/structure alignment

aln.align2d(max_gap_length=40)

# Write out resulting alignment in both PIR and PAP formats

aln.write(file='TvLDH-1Y7tA.ali',alignment_format='PIR')

aln.write(file='TvLDH-1Y7tA.pap',alignment_format='PAP')
```

文件输入 automodel，给出代码和所建模型数量。

```
from modeller import *
from modeller.automodel import *
env = environ( )
env.io.atom_files_directory=['.','../atom_files']
a – automodel(env,alnfile='../step3_align/TvLDH1y7tA.ali',)
knowns='1y7tA',sequence='TvLDH',assess_methods=assess.DOPE)
a.starting_model = 1
a.ending_model = 5
a.make( )
```

（7）通过 MODELLER 中的 Mod – EM 将 GPR116 的 cryo – EM 数据导入，使目标蛋白结构符合这些参数。

```
from modeller import *
log.verbose( )
env = environ( )
struct='../step4_model/TvLDH.B99990001.pdb'
map='TvLDH.10A.mrc'
resoluton=10.0
box_size=48
apix=1.88
x=–26.742;y=–9.5205;z=–10.375 #origin
steps=20
# Read in cryo=EM density map
den = density(env,file=map,em_density_format='MRC',voxel_size=apix,resolution=resolution,
em_map_size=box_size,density_type='GAUSS',px=x,py=y,pz=z)
# Fit the PDB file into the map by MC simulated annealing
den.grid_search(em_density_format='MRC',num_structures=1,em_pdb_name=struct,chains_num=[1],
start_type='CENTER',number_of_steps=steps,angular_step_size=30.,temperature=100,
best_docked_models=1,translate_type='RANDOM',em_fit_output_file='modem.log')
```

（8）目标模型的评价。通过上述步骤得到最好的模型，使用 MODELLER 标准评分函数 molecule PDF（Molpdf）方法进行评分，结果分别在 PDB 开头的 REMARK 中赋值。此外，还有 DOPE 和 GA341（概率分布函数）评价方法。对于一个可信度高的模型，其 GA341 值越接近于 1；当 GA341 值小于 0.6 时，认为该模型可信度低，不能反映该蛋白的立体构象。而 Molpdf 和 DOPE 则代表该蛋白模型的"能量"，因此，在 GA341 值接近于 1 的情况下，Molpdf 和 DOPE 值越小越好。

2. GPR116 的 MODELLER 建模结果

目标蛋白 GPR116 的立体模型构建完成后，共得到 5 个模型（表 5 – 4）。运行 evaluate_model.py 脚本，对这 5 个模型进行质量评价。由表 5 – 4 可以看出，第一个模型 GA341 值最接近于 1，且不小于 0.6；Molpdf 值最小，DOPE 值也较小。因此，选择 GPR116.B99990001.pdb 作为最佳模型用于分子对接。

表 5 – 4　GPR116 模型质量评价

Table 5 – 4　GPR116 model quality estimate

Filename	Molpdf	DOPE Score	GA341 Score
GPR116.B99990001.pdb	1 401.59778	– 24102.40430	0.95528
GPR116.B99990002.pdb	1 495.80579	– 24555.36914	0.81127
GPR116.B99990003.pdb	1 512.26050	– 24068.83398	0.84483
GPR116.B99990004.pdb	1 514.73669	– 24021.78125	0.82945
GPR116.B99990005.pdb	1 465.48267	– 24180.48828	0.72811

其他 25 个 G 蛋白偶联受体所构建模型采用同样的评分函数进行评价，同样的方法选择最佳模型。

四、讨论

随着蛋白质组学的深入研究，目前已获得大量蛋白质的晶体结构。解析蛋白质的立体空间结构是一个耗时昂贵的过程，目前对蛋白质晶体结构的解析方法在一定程度上都存在限制或缺陷。X – ray 单晶体衍射技术的明显缺陷是必须获得蛋白质晶体，但许多蛋白质不能结晶，其应用范围受到了限制；核磁共振技术（Nuclear magnetic resonance，NMR）

第五章
G 蛋白偶联受体的筛选及同源建模

不需要蛋白质结晶,但需要高浓度的蛋白质溶液,且蛋白质稳定、不变性、不发生聚集,尤其是 NMR 技术不能解析大分子蛋白质结构。电镜三维重构技术要求获得蛋白质二位晶体,但目前能够获得的结构分辨率较低,因此,该技术也不能作为常规方法测定蛋白质。针对这些问题,目前获得蛋白质立体结构的捷径就是依靠同源模建(Homology modeling)来预测,该方法的预测基础是:氨基酸序列的同源性决定了三维立体结构的同源性。研究发现,目标蛋白序列与模板蛋白序列同源性在 50% 以上时,获得的目标蛋白模型可信度很高;若两蛋白同源性在 30%~50% 时,获得的目标蛋白模型可信度次之;若两蛋白同源性在 30% 以下时,获得的目标蛋白模型可信度较差。同源建模的一般步骤为:同源蛋白搜索、序列比对、模型建立与优化以及模型评价四部分。

目前,蛋白质三维结构预测有 3 种方法:折叠识别法、从头预测法和同源建模法。同源建模又称比较建模,是当今最实用和可信度高的蛋白结构预测方法。

Browne 等(1969)世界上首次以蛋清溶菌酶结构为基础,成功对 α-乳清蛋白三维结构手工模建。从 Blundell 等(1987—1988)正式提出同源蛋白三维结构预测之后,相继出现了大量通过同源建模、参数设定条件下成功预测各种蛋白三维结构的报道。蛋白质同源建模的理论基础就是依据生物在进化过程中,蛋白质立体空间结构保守性远大于氨基酸序列保守性,因此,当蛋白质之间的同源性大于 30% 时,其空间骨架结构基本不发生变化。

本章内容是以 195 个膜蛋白为对象,通过 HMMTOP V2.0 对跨膜 α 螺旋进行了分析,获得 38 个 G 蛋白偶联受体。然后利用 SWISS-MODEL 和 MODELLER 同源建模技术对 CART 和 38 个 G 蛋白偶联受体进行分子建模,并利用分子动力学的能量最小化原理反复优化,经 QMEAN4 评分函数(SWISS-MODEL)、标准评分函数 Molpdf、DOPE 和 GA341(MODELLER)验证,所构建的 39 个蛋白的分子模型结构合理真实,为深入研究筛选 CART 受体奠定基础。

小结

1. 通过 HMMTOP 对 195 个膜蛋白进行了跨膜 α 螺旋分析，获得 38 个 G 蛋白偶联受体。

2. 通过 SWISS–MODEL 获得配体 CART 和 12 个 G 蛋白偶联受体的空间结构，并成功建立分子坐标。

3. 通过 MODELLER 建模软件获得其余 26 个 G 蛋白偶联受体的可信合理的空间结构，并成功建立分子坐标。

第六章

CART 与筛选蛋白的分子对接研究

一、概述

蛋白质与蛋白质分子对接技术是通过高效的计算机模拟技术研究蛋白质—蛋白质间相互作用与识别的有效方法,该技术不仅有助于揭示蛋白分子间的识别原理和蛋白质序列一级结构、二级结构、三级结构与功能的关系,同时也带动其他相关领域的应用研究,如信号调节、药物设计、抗原与抗体、酶与底物等蛋白质工程分子设计方面的研究。

ZDOCK 分子对接要求提供配体和受体的三维分子坐标,其主要步骤如下:①通过调整分子间距离和原子旋转角度搜集配体—受体的各种结合模式;②通过各种设定参数,将一些明显不能结合的数据过滤掉;③从其他的对接模式中筛选出最接近复合物天然构象的结合模式或蛋白。从热力学角度分析,蛋白质分子间的相互作用和识别过程是一个热力学平衡的过程,二者形成的稳定复合物,其结合自由能处于最低状态。因此,通过几何互补、去溶剂化能和静电相互作用评分函数评价对接分子的结合自由能,进而筛选 CART 受体及二者最佳的结合模式,这无疑是最直接的方式。

DOCK 是基于快速傅里叶转换(Fast fourier transform, FFT)的分子对接算法,对一对大分子蛋白质的对接结构通过几何互补、去溶剂化能和静电相互作用进行评分函数筛选,是目前蛋白质—蛋白质分子对接中的通用程序。研究中利用 CART(做为配体)和 38 个 G 蛋白偶联受体（作为受体）的 PDB 文件作为材料，在 ZDOCK V3.0.2 提交配体/受体 PDB 文件，建立受体配体结合中心网格，设定蛋白分子结合距离＜6Å，预测数量 1 000，评分函数值由高到低，显示结合构象为前五个预测结果，其他参数软件自动默认，然后进行分子空间对接。结果显示：38 个分子对接结果中，复合体最高评分范围为 1 442.716～2 125.649。根据评分结果，将 TEDDM1、CMKLR1、AGTR2 和 GPR16 作为 CART 的候选受体。

二、材料与方法

（一）供试材料

本试验以 CART 作为配体，提供 CART 的 PDBID 代码；以 38 个 G 蛋白偶联受体作为 CART 的候选受体，分别提供 CHRM5、GPR39、APLNR、LPAR6、CCBP2、AGTR2、S1PR4、RRH、CASR、ADRA1B、LGR5 和 CMKLR1，共 12 个受体的 PDBID 代码或 PDBQT 文件；分别提供 GPR116、MRGPRF、GPR4、PQLC3、LOC782475、GPR124、GPER、MBOAT4、P2RY6、FZD4、SRD5A3、GPR137C、LOC526047、LOC514434、EMR1、LOC782554、LOC524771、GPR156、TEDDM1、LOC787041、GPR142、APH1B、LOC614050、GPR85、FZD3 和 LOC512627，共 26 个受体的 PDBQT 文件。

（二）计算平台及软件

1. 硬件平台

进行 ZDOCK 分子对接时，对计算机硬件环境的要求主要是：CPU（四核 Intel 及以上）；内存（4.00 G 及以上）；显卡（NVIDIA Quadro NVS 285）。其他随计算机配置。

2. 软件平台

64 位中文版 Lunix 操作系统。

第六章
CART 与筛选蛋白的分子对接研究

Java SE Development Kit（JDK7）V7.0 用于 JMol（分子形象化展示）。

ZDOCK V3.0.2（http：//zdock.umassmed.edu/）：ZDOCK V3.0.2 服务平台由美国马萨诸塞大学药学院开发与维护，主要用于大分子蛋白质间的分子对接。该网站界面由 PHP 和 HTML 语言写入，是基于对象和事件驱动的客户端脚本运行，能完美展示由分子组成的可视化、形象化的空间构象。

（三）试验方法

1. 输入结构和选择项

进入 ZDOCK SERVER 界面（http：//zdock.umassmed.edu/），选择 ZDOCK 选项。在"Input Protein 1"输入受体蛋白 PDBID 代码或 PDBQT 文件，显示蛋白描述信息，选择"使用生物学装配"；在"Input Protein 2"输入 CART 的 PDBID 代码，显示蛋白描述信息，选择"使用生物学装配"。选择 ZDOCK 版本 3.0.2，不选择残基跳跃，然后提交输入结果，见图 6-1A 所示。

2. 选择网格与活性位点

这一步主要根据受体、配体的结构特征和结合部位，分别选择受体、配体的模拟结合网格和结合位点，即通过选择序列中的某段氨基酸或某几个特定氨基酸残基来实现，如图 6-1B 所示，左右两边分别为提交的受体、配体的立体空间结构。ZDOCK 自动默认参数设定为：蛋白分子结合距离 <6Å，预测数量 1 000，评分函数值由高到低，显示结合构象为前 5 个预测结果。然后提交，ZDOCK 平台随后邮件确认，10~20min 将分析结果以网页链接发至邮箱。

3. 查看结果

收到的分析页面链接是计算机随意合成的一段编码，目的是保护使用者的分析信息。在页面上方提供分析结果输出、使用者预先输入的配体、受体 PDB 文件下载以及前 10 个预测结果的下载（图 6-1C）。结

牛卵泡可卡因－
苯丙胺调节转录肽（CART）受体的筛选

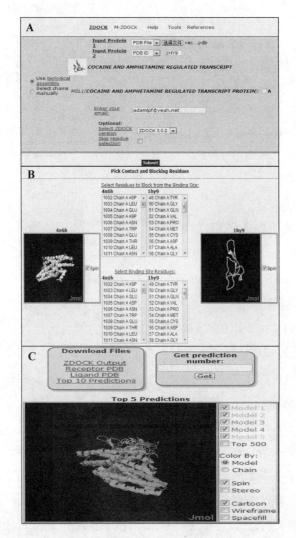

图 6-1　ZDOCK 服务平台：A. PDBID/PDBQT 输入，
B. 网格和结合位点设定，C. 输出结果

Figure 6-1　The ZDOCK Server interface：A. Structure Input,
B. Blocking/Contacting Residue Selection, and C. Results pages.

果页面的特点是有一通过 JAVA 程序运行的 JMol 可视化图像，该图像可以展示前 500 个预测结果的立体结合构象。

第六章
CART 与筛选蛋白的分子对接研究

三、结果与分析

（一）蛋白复合体立体构象模型

将获得的链接打开,即可获得所输入配体-受体的结合图像（图6-1C）,图中显示的依次是得分最高的前5个结合模型,模型可以线性模型、框架模型和立体空间模型3种方式显示。保留得分最高的第一个复合体模型,选择立体空间模型。CART 与38个 G 蛋白偶联受体形成复合体的立体空间模型如图6-2所示。

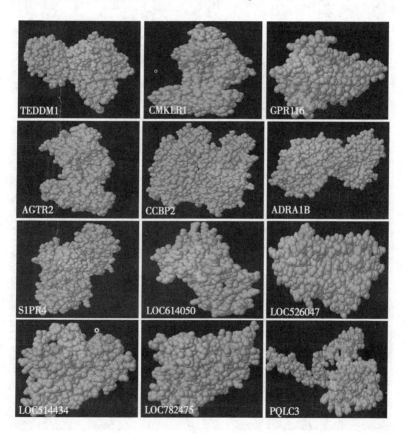

牛卵泡可卡因－
苯丙胺调节转录肽（CART）受体的筛选

（续图6-2）

P2RY6　LOC524771　LOC782554
MRGPRF　MBO AT4　LPAR6
LOC512627　LOC787041　LGR5
GPR156　GPR137C　RRH
GPR85　GPR142　GPR39
GPER　GPR4　GPR124

（续图6-2）

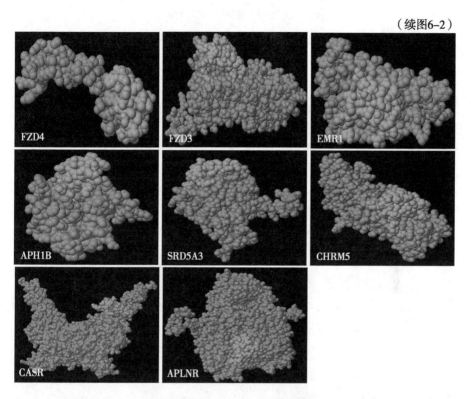

图 6-2　蛋白复合体立体空间模型

Figure 6-2　Protein complex three-dimensional space model.

（二）评分结果输出

分子对接后，在结果输出界面，点击"ZDOCK Output"可获得1 000个在"Grid"内对接结果的评分，选择最大评分作为两蛋白结合的最佳模型（表6-1）。其中，"Grid"为两蛋白对接时，选定的"受体"结合部位的空间网格大小，如CMKLR1在对接时，Grid=140代表参数设定时，该蛋白结合部位的立体空间大小为140×140×140，每个网格的大小为1.2Å，也即该网格大小为边长168Å的立方体。1 000个对接结果来自于"配体"以不同的立体结构在该范围内与"受体"形成的复合体。

牛卵泡可卡因－苯丙胺调节转录肽（CART）受体的筛选

表 6–1　分子对接最高评分输出

Table 6–1　ZDOCK Max score output

Protein Name	PDBQT/PDBID	Grid	Max Score
CMKLR1	PDBID	140	1 977.343
GPR156	PDBQT	128	1 577.257
APLNR	PDBID	140	1 757.454
MRGPRF	PDBQT	108	1 775.029
GPR4	PDBQT	112	1 802.961
MBOAT4	PDBQT	180	1 696.818
LPAR6	PDBID	140	1 660.509
GPR116	PDBQT	108	1 960.165
CCBP2	PDBQT	128	1 805.442
LGR5	PDBID	140	1 558.230
GPR85	PDBQT	112	1 575.595
GPR39	PDBID	128	1 720.169
CHRM5	PDBID	128	1 545.457
PQLC3	PDBQT	168	1 555.330
LOC782475	PDBQT	108	1 826.363
GPR124	PDBQT	100	1 688.348
GPER	PDBQT	112	1 707.954
AGTR2	PDBID	140	1 977.343
P2RY6	PDBQT	112	1 475.663
FZD4	PDBQT	112	1 685.837
SRD5A3	PDBQT	108	1 915.821
GPR137C	PDBQT	180	1 655.180
LOC526047	PDBQT	108	1 509.955
LOC514434	PDBQT	112	1 740.311
EMR1	PDBQT	100	1 680.506
LOC782554	PDBQT	112	1 782.226
LOC524771	PDBQT	128	1 893.939
S1PR4	PDBID	140	1 672.789
TEDDM1	PDBQT	128	2 125.649
LOC787041	PDBQT	112	1 712.800
GPR142	PDBQT	108	1 723.823
LOC512627	PDBQT	112	1 907.312
LOC614050	PDBQT	108	1 599.558

(续表)

Protein Name	PDBQT/PDBID	Grid	Max Score
RRH	PDBID	192	1 919.800
CASR	PDBID	140	1 442.716
ADRA1B	PDBID	140	1 764.445
FZD3	PDBQT	128	1 825.801
APH1B	PDBQT	108	1 807.397

通过输出结果的最高评分，选择 TEDDM1，CMKLR1，AGTR2 和 GPR116 4 个蛋白作为 CART 的候选受体进行后续研究。

四、讨论

ZDOCK 是目前大分子蛋白质对接运算中最突出的程序，也是基于 FFT 的方法。ZDOCK 与其他基于 FFT 对接运算的共同特征是：ZDOCK 也是将静电能量、几何匹配的运算途径通过求卷积的方法来求解，应用 FFT 方法加速；不同点在于 ZDOCK 使用基于几何匹配的打分方式 PSC（Pairwise shape complementarity）和基于原子配对统计的打分方式 ACE（Atomic contact energy）。

PSC 评分方式与普通几何匹配 GSC 的根本不同在于：GSC 将蛋白表面格点均复制为 1，当配体—受体距离靠近时，每一个接触的格点均得 1 分；而 PSC 将受体不同表面的格点赋予不同的数值，也即 PSC 增大了配体—受体结合位点（沟槽）的表面赋值。当受体存在较大较深的"沟槽"时，PSC 就偏倾向于在"沟槽"处采样得分。这种评分方式与配体—受体的真实结合模式是相吻合的，大多数配体—受体形成的蛋白复合体都是一个蛋白的突出部分与另一蛋白"沟槽"相结合。因此，分子对接中 PSC 评分方法的预测结果明显优于其他普通的几何匹配。

ACE 评分方式是 Zhang 等（1997）基于统计学原理，对 89 个蛋白

牛卵泡可卡因 — 苯丙胺调节转录肽（CART）受体的筛选

质不同原子之间形成的 contact 能量进行计算，获得原子结合能模型，即 ACE。ZDOCK 的 FFT 采样环节中运用该统计势，提高了 ZDOCK 分子对接结果的可信度和成功率。在本章研究内容，ZDOCK 3.0.2 中，ACE 被优化成 12 种原子。通过 FFT 变换完成对这 12 种原子之间的结合能计算。因此，PSC 评分方式，加上库仑势并结合 ACE，使得 ZDOCK 预测配体—受体成功率优于其他基于 FFT 方法。通过 ZDOCK 分子对接技术获得每个蛋白复合体的最高评分，从而选择 TEDDM1，CMKLR1，AGTR2 和 GPR116 4 个 G 蛋白偶联受体作为 CART 的候选受体。

受体就是介导细胞信号转导的一类功能蛋白，通过识别细胞内外环境中的某些微量物质并与其结合，从而使信号系统放大并触发相关生理反应。受体与配体的特定结合部位称为受点（Receptor site）。受体按照在细胞中位置的不同可分为：细胞膜受体、胞浆受体和细胞核受体；也可按受体本身的特征大致分为：离子通道受体、调节基因表达的受体、G 蛋白偶联受体和具有酪氨酸激酶活性的受体。作为受体必须满足以下条件：①特异性，即细胞特异性和配体识别特异性；②亲和性，即受体—配体结合具有高度的亲和性；③饱和性，即配体可以饱和受体，否则无法实现复杂调控机制；④有效性，即受体—配体的结合会诱发某种生物学效应。

目前研究认为，G 蛋白偶联受体属于一类膜受体的统称，其特点是空间结构中均有 7 个跨膜 α 螺旋，该结构域将受体蛋白分割为膜内 C 端、膜外 N 端、3 个膜内 Loop 环和 3 个膜外 Loop 环；同时，在肽链 C 端和膜内 Loop 环上均有 G 蛋白（鸟苷酸结合蛋白）结合位点。迄今为止，对 G 蛋白偶联受体与配体结合后蛋白构象变化的研究还不是很清楚，通常认为，配体是与非激活状态的 G 蛋白偶联受体发生结合，而后发生构象变化，激活 G 蛋白调节下游反应。

Rasmussen 等（2007）首次揭示了第一个人类 G 蛋白偶联受体——β2 肾上腺素能受体的结构，之后在 2011 年获得了 G 蛋白与 β2 肾上腺

第六章
CART 与筛选蛋白的分子对接研究

素能受体复合体的晶体结构。2012 年《Nature》上又公布了一个 G 蛋白偶联受体家族蛋白——趋化因子受体 1（Chemokine receptor 1, CXCR1）的晶体结构被解析。CXCR1 为白细胞介素的膜受体，对人体炎症反应和免疫反应具有重要作用，虽然对 CXCR1 下游信号通路还不明确，但在获得晶体结构的基础上更容易进行功能的深入探讨。传统意义上，由于 G 蛋白偶联受体的空间结构在细胞膜内外都有较长的 Loop 环存在，因此，造成其立体结构的稳定性很差，通过 X－射线衍射法获得其晶体结构时，必须通过剪切和修饰使其结构的稳定性提高。在应用型研究方面，通常 G 蛋白偶联受体作为许多药物设计的靶点来发挥药物的功效，使得 G 蛋白偶联受体具有广泛的应用前景。

TEDDM1（Transmembrane epididymal protein 1）为跨膜附睾蛋白 1。对牛全基因组研究发现该基因全长 1 415 bp，编码 509 个氨基酸。Yamazaki 等（2006）通过基因芯片技术发现 *TEDDM*1 在小鼠附睾中高表达；阉割后的小鼠附睾中该基因表达量降低，注射睾酮也不能恢复其高表达，E_2 的分泌也不可恢复。目前，对于该蛋白功能的研究仍属于空白，有待于研究者深入探讨。

CMKLR1（Chemokine - like receptor 1）为趋化因子样受体 1，在人类研究中也称 ChemR23，属于分泌型蛋白。*CMKLR*1 在动物体内广泛表达，尤其在心肌细胞、巨噬细胞、内皮细胞、浆细胞样树突状细胞、脂肪细胞中显著高表达。*CMKLR*1 全长 1 283 bp，CDS 区 1 089 bp，编码 362 个氨基酸。属于 G 蛋白偶联受体（GPCR）家族成员。Arita 等研究还发现，*CMKLR*1 在具有免疫及造血功能的组织中广泛表达，如骨髓、脾脏、胎儿肝脏、淋巴结、胸腺和阑尾。1996 年 Owman 等从人 B 淋巴母细胞 cDNA 文库中鉴定出一段与 G 蛋白偶联受体家族（GPCRs）高度同源的 cDNA 序列，将其命名为 *CMKLR*1，后来研究证实为 Chemerin 脂肪因子的天然受体。目前，对 CMKLR1 生物学功能的研究主要有以下几方面：①参与免疫与炎性反应。当动物机体出现炎性反应时，巨噬细胞和树突状细胞发挥其抗原递呈作用，向炎性反应部位集合，通过

牛卵泡可卡因-苯丙胺调节转录肽（CART）受体的筛选

CMKLR1 与其配体 Chemerin 相互作用，诱导胞内钙离子释放和磷酸化，同时胞外信号调节激酶 ERK1/2 通过结合 Gi-偶联的异源三聚体来抑制 cAMP 的累积，从而完成对炎性部位的免疫反应。目前认为，肥胖属于慢性低水平的炎性状态，可引起脂肪组织炎性反应。Cash 等（2008）通过研究小鼠脂肪细胞的抗炎特性发现，其抗炎机理可能是 Chemerin 通过募集表达有 CMKLR1 的巨噬细胞和浆细胞样树突状细胞聚集，从而激活免疫细胞发挥抗炎特性。②调节脂肪细胞分化。Goralski 等（2007）通过对沙鼠的研究发现，在沙鼠脂肪细胞分化的早期阶段和脂肪细胞成熟时，*CMKLR*1 表达显著增加；通过基因沉默技术进一步研究发现，CMKLR1 在脂肪细胞分化早期（前 3d）具有重要作用，但在第 4d 将 CMKLR1 基因沉默，则对脂肪细胞分化和脂肪细胞表型无明显影响。③参与胰岛素抵抗。Takahashi 等通过培养小鼠 3T3-L1 细胞研究发现，CMKLR1 与其配体 Chemerin 相互作用，可显著增加胰岛素刺激引起的葡萄糖摄入，并能提高 ISR-1（胰岛素受体底物-1）酪氨酸的磷酸化水平，增强胰岛素刺激信号。

AGTR2（Angiotensin II receptor, type 2）为血管紧张素 II 受体-2 型，该基因编码 362 个氨基酸。目前，AGTR2 的生物学功能还不是很清楚，研究的方向主要集中在：AGTR2 可能对心脏、血管起保护作用、可能与大脑发育和认知功能有关，AGTR2 也可能对 AGTR1 起负调控作用，参与抑制细胞生长、促进细胞凋亡的生物学过程。

GPR116（G protein-coupled receptor 116）为 G 蛋白偶联受体 116，最初在小鼠的肺和肾中发现，具有和 IG 相似的结构特点。人类 GPR116 基因定位于 6p12.3 染色体上，编码 1 346 个氨基酸。目前，对 GPR116 的研究基本属于空白，没有已知的配体，没有信号通路的相关研究，对其功能的研究仅有一篇报道。Tang 等（2013）通过对 Adhesion G 蛋白偶联受体家族进行表达和功能筛查，发现 GPR116 属于一个新的乳腺癌转移调控因子；进一步研究发现在高度转移性的乳腺癌细胞中，下调 GPR116 表达可以抑制细胞迁移和侵袭；反之，在转移性较差的乳腺癌

第六章
CART 与筛选蛋白的分子对接研究

细胞中异位表达 GPR116 能促进细胞侵袭。此外，研究者通过两个乳腺肿瘤转移小鼠模型证实，下调 GPR116 可以抑制癌细胞转移。

小结

1. 通过 ZDOCK 3.0.2 对 CART 和 38 个 G 蛋白偶联受体进行分子对接，获得 38 个复合体立体空间结构。

2. 通过 FFT 方法加速，应用几何匹配打分方式 PSC 和原子配对统计打分方式 ACE 对复合体结构进行评分，选择 TEDDM1，CMKLR1，AGTR2 和 GPR116 4 个 G 蛋白偶联受体作为 CART 的候选受体。

第七章

CART 及其候选受体基因在牛优势卵泡和从属卵泡中的表达

第一节 牛发情周期第一卵泡波优势卵泡和从属卵泡的鉴定

一、概述

所谓优势卵泡就是在动物卵泡发育过程中，1个卵泡阻止其他卵泡的生长或1个卵泡生长的激素环境不适宜其他卵泡的生长。根据这个定义，通过超声波检测，优势的起始就是发情周期内：①同一个卵泡波内优势卵泡的直径至少大于从属卵泡1~2mm的时间；②同一卵泡波中所有从属卵泡停止生长的时间。根据定义可知通过超声波检测或生物化学分析，鉴定优势卵泡是可能的，但只有当所有从属卵泡生长停止才能确定。优势的丧失可通过超声波来检测，与下一个卵泡波中发生出现的时间相一致。从属卵泡指在一个卵泡波中，通过超声波检测确认了优势卵泡以后剩余的卵泡，牛通常为2~5个。

1984年以前，研究牛发情周期内卵泡生长和闭锁的动力学困难重重，主要由于对牛卵泡的形态学观察方法单一，通过屠宰牛或切除卵巢后获得卵泡进行研究，尤其是对卵泡的发育过程进行研究至少需要两次外科手术。1984年，有两个课题组报道通过一维数组超声波扫描仪对

第七章
CART 及其候选受体基因在牛优势卵泡和从属卵泡中的表达

奶牛和小母牛的卵巢结构进行检查。然而，Pierson 和 Ginther 是首次通过超声波扫描术来检测牛整个发情周期卵泡的直径。超声波扫描术所获得的卵泡直径通常是指卵泡无反射波的空腔的宽度，里面充满卵泡液，因此，通过超声波扫描术获得的卵泡直径要比同一卵泡通过外科手术获得的直径小 2~3mm。

1982 年，Ireland 和 Roche 通过对卵泡液中雌激素与孕酮或雄激素的比值首次对牛有腔卵泡（直径≥6mm）进行分类，分为雌激素活跃的卵泡和雌激素不活跃的卵泡。雌激素活跃的卵泡（雌激素浓度＞孕酮和雄激素浓度）具有健康生长卵泡的生物化学特性，雌激素不活跃的卵泡有闭锁卵泡或注定向闭锁卵泡发育的特性。本试验通过采集牛卵巢，通过对黄体内外颜色形态、黄体直径、表面脉管系统及卵泡直径的观察，确定所采集的卵泡处于牛发情周期的第一卵泡波；从处于第一卵泡波的牛卵巢上分离最大卵泡和第二大卵泡，抽取卵泡液，采用牛 E_2 Elisa Kit 试剂盒、PROG Elisa Kit 试剂盒分别进行 E_2 和 P 的测定，通过大、小卵泡 E_2 和 P 的比值确定所获得的卵泡是否为 DF 和 SF。

二、材料与方法

（一）试验动物及其组织

本试验材料均来自山西省文水县胡兰镇贤美肉牛屠宰场。母牛（西门塔尔牛杂交后代）屠宰后采集双侧卵巢，编号后投入冰盒中制备好的杜氏磷酸缓冲液（DPBS）中，迅速带回实验室。

（二）试验器材

液氮、无 RNase EP 管、无 RNase 枪头、PE 手套、橡胶手套、口罩、研钵、眼科剪和镊子等。

（三）试验方法

1. 卵泡液的分离

牛卵泡可卡因 — 苯丙胺调节转录肽（CART）受体的筛选

（1）取卵泡。在细胞培养室将盛 70% 酒精和 DPBS 溶液两个小培养皿放于纸巾上，一小烧杯内盛少量的培养液放于冰上。通过对黄体观察，选择处于第一卵泡波的卵巢，在一对卵巢中选择最大卵泡、第二大卵泡，用眼科手术剪轻轻从卵巢上剪下，并将剪下的卵泡在 70% 酒精和 DPBS 溶液中各浸数秒后用镊子夹入盛培养液的烧杯内。

（2）抽取卵泡液。卵泡转移到冰上新鲜的培养液中。在超净台中，卵泡放置到无菌的表面皿上，使用无菌的 3ml 注射器及 22g 针头，预先湿润后，轻轻注入卵泡内，缓慢地分别抽取各组卵泡液，并将卵泡液放置在适当的带有标签的试管中储存（每个卵泡使用不同的注射器和针头），-20℃ 保存备用。然后将卵泡用无菌的冰冷培养液洗涤，清洗液用移液管吸走丢弃。表面皿中加入新鲜的培养液用于分离颗粒细胞。

2. 雌激素和黄体酮的测定

本试验牛卵泡颗粒细胞 E_2 和 P 的测定采用竞争法，用牛 E_2 Elisa Kit 试剂盒和 PROG Elisa Kit 试剂盒进行测定，试验方法参照说明书。

三、结果与分析

（一）通过黄体特征确定第一卵泡波

在采样过程中，共获得 11 对卵巢的黄体形态符合第一卵泡波黄体形态的要求（图 7-1 对照组）。第一卵泡波黄体形态为：从黄体的外部观察，黄体呈红色，说明卵巢近日排卵，同时排卵裂点未被上皮细胞覆盖；黄体直径在 0.5~1.6cm；双侧卵巢上最大卵泡直径 ≤13mm；黄体表面的脉管系统不明显；手术刀将黄体切开后，内部呈红色，偶尔充满血液，细胞松弛且排列有规则（图 7-1）。

（二）卵泡液雌激素和孕激素的测定

利用 E_2 Elisa Kit 试剂盒和 PROG Elisa Kit 试剂盒、酶标仪对各组最大卵泡、第二大卵泡卵泡液进行 E_2、P 的测定（表 7-1），结果显示：

第七章
CART及其候选受体基因在牛优势卵泡和从属卵泡中的表达

第3组、第5组和第7组卵巢卵泡液 E_2 和 P 测定数值中，最大卵泡的 E_2/P 值 >1，第二大卵泡的 E_2/P 值 < 1，符合 DF、SF 的筛选要求。因此，第3组、第5组和第7组卵巢采集的卵泡为第一卵泡波中的 DF 和 SF。

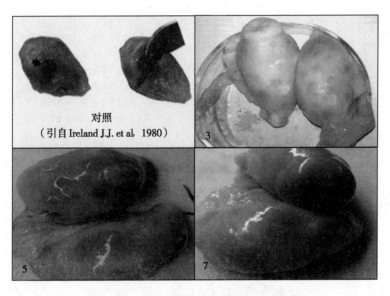

图7-1 牛发情周期第一卵泡波黄体形态

Figure 7-1 Corpus luteum morphological feature of the first follicular wave during the estrous cycle of cattle

表7-1 各组卵巢卵泡液 E_2 和 P 的测定结果

Table 7-1 The E_2 and P determination results from groups of ovarian follicle fluid

编号	卵泡类型		激素测定值（ng/ml）				平均值	E_2/P
1	最大卵泡	E_2	2.8599	3.3538	2.7101	3.2893	3.0533	0.2760
		P	10.9704	10.9704	11.0938	11.2198	11.0636	
	第二大卵泡	E_2	13.0440	13.3508	14.0221	14.0221	13.6098	1.2438
		P	10.7308	10.8494	11.0938	11.0938	10.9420	
2	最大卵泡	E_2	3.2735	3.1510	2.9660	2.8341	3.0562	0.2786
		P	10.9704	10.8494	10.9704	11.0938	10.9710	
	第二大卵泡	E_2	11.7367	10.7100	11.7367	11.0935	11.3192	1.0143
		P	10.9704	10.9704	11.3484	11.3484	11.1594	

牛卵泡可卡因-苯丙胺调节转录肽（CART）受体的筛选

（续表）

编号	卵泡类型		激素测定值（ng/ml）					平均值	E_2/P
3	最大卵泡	E_2	17.8867	19.3434	20.1940	19.3434	19.1919		1.9668
		P	10.2786	9.2829	9.1925	10.2786	9.7582		
	第二大卵泡	E_2	4.2035	4.3017	4.7714	4.7411	4.5044		0.4356
		P	9.6603	9.7588	10.9704	10.9704	10.3400		
4	最大卵泡	E_2	11.5123	11.0935	10.1908	10.3570	10.7884		0.9777
		P	11.2198	11.2198	10.8494	10.8494	11.0346		
	第二大卵泡	E_2	3.0494	3.0212	2.6045	2.5704	2.8114		0.2606
		P	10.8494	10.8494	10.7308	10.7308	10.7901		
5	最大卵泡	E_2	12.2194	12.4797	12.7541	13.3508	12.7010		1.1738
		P	10.7308	10.7308	10.9704	10.8494	10.8204		
	第二大卵泡	E_2	3.0212	3.1069	3.2578	3.1810	3.1417		0.2975
		P	10.7308	10.7308	10.3885	10.3885	10.5597		
6	最大卵泡	E_2	8.3842	8.5957	7.2544	7.4835	7.9295		0.8929
		P	7.1467	7.0287	10.7308	10.6145	8.8802		
	第二大卵泡	E_2	7.3290	7.1815	7.6457	7.6457	7.4505		0.7429
		P	10.7308	10.0650	9.7588	9.5635	10.0295		
7	最大卵泡	E_2	13.6762	13.3508	14.7846	15.2067	14.2546		1.3207
		P	11.0938	10.8494	10.7308	10.5004	10.7936		
	第二大卵泡	E_2	3.2422	3.2893	3.4544	3.3703	3.3391		0.4121
		P	9.3749	9.3749	6.7462	6.9136	8.1024		
8	最大卵泡	E_2	10.1908	10.7100	11.5123	11.2982	10.9278		0.9209
		P	12.9646	12.8003	10.9704	10.7308	11.8665		
	第二大卵泡	E_2	4.6816	4.6235	4.2521	4.2277	4.4462		0.3528
		P	12.6398	12.4831	12.8003	12.4831	12.6016		
9	最大卵泡	E_2	15.6604	16.1500	15.6604	15.2067	15.6694		0.8556
		P	18.3146	18.3146	18.3146	18.3146	18.3146		
	第二大卵泡	E_2	5.3142	5.3523	4.4571	4.4841	4.9019		0.3191
		P	15.5368	15.5368	15.3044	15.0787	15.3642		
10	最大卵泡	E_2	15.2067	15.2067	14.7846	15.6604	15.2146		0.8038
		P	19.3547	19.7288	18.3146	18.3146	18.9282		
	第二大卵泡	E_2	3.4544	3.3053	3.1510	3.0494	3.2400		0.2346
		P	14.2375	14.0413	13.4825	13.4825	13.8110		
11	最大卵泡	E_2	3.6707	3.4203	2.9797	2.9525	3.2558		0.3285
		P	9.8591	9.7588	10.0650	9.9611	9.9110		
	第二大卵泡	E_2	11.7367	11.9721	11.9721	11.2982	11.7448		0.7126
		P	16.2776	16.0230	16.8117	16.8117	16.4810		

第七章
CART 及其候选受体基因在牛优势卵泡和从属卵泡中的表达

四、讨论

（一）依据黄体形态特征确定第一卵泡波

对牛发情周期内卵泡波所处阶段的确定，通用的方法有墨汁标记法、超声波扫描法和黄体形态观察法。墨汁标记法对卵泡发育过程进行研究，至少需要两次外科手术，对牛机体及卵巢发育影响较大，严重会导致卵泡发育受阻，因此，这种方法目前已很少使用。超声波扫描术所获得的卵泡直径通常是指卵泡无反射波空腔的宽度，里面充满卵泡液，因此，通过超声波扫描术获得的卵泡直径要比同一卵泡通过外科手术获得的直径小 2~3mm。同时，超声波扫描监测中，需对母牛整个发情期每天 2~3 次超声波检测并绘制卵巢卵泡发育图，操作上较为困难且成本较高。因此，在本实验中，通过在肉牛屠宰场大量采集母牛卵巢，应用黄体形态观察法仔细进行鉴别，并成功获得 11 对符合要求的卵巢。

McNutt 和 Hammond 报道在牛发情周期内黄体的形态上有明显的变化，Ireland 等（1980）通过对黄体表观总的变化特征的研究，提出了准确判断母牛发情阶段的方法。在 Ireland 等的研究中，选择未知年龄海福特小母牛 180 头作为研究对象，按遗传和品种的分类标准每 6 头分一栏，共 30 栏。动物每周称体重并用声纳射线测定背膘厚度，当每栏牛的平均背膘厚度达 12mm 后进行屠宰，其中，实验日粮对生殖现象无影响。在生长试验中，每天 2 次观测母牛发情行为（7:00~8:00，16:00~17:00）。同时通过对母牛的爬跨行为和卡马氏发情测定器确定发情状况。在屠宰时，颈静脉采血并用放免法分析孕酮浓度。通过对各母牛黄体的形态特征进行研究，将这些特征分成 4 个阶段，通过与已知的发情周期的天数对比，确定了准确判断发情周期 4 个阶段的方法。

第Ⅰ阶段（发情周期第 1~4d）包括排卵和上皮细胞生长覆盖至裂点之间的时间（图 7-2，Ⅰ图）。形成一个新黄体的顶点，标志着第Ⅱ阶段开始；第Ⅱ阶段（发情周期第 5~10d）黄体外周由明显的脉管系

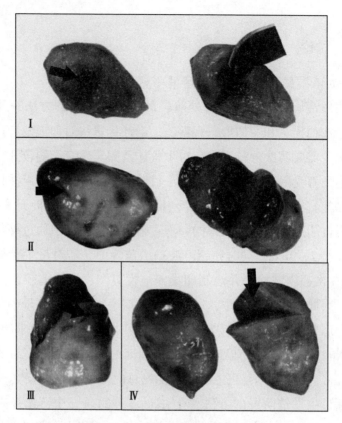

图 7-2 发情周期各阶段黄体的变化

（引自 James J Ireland. et al., 1980）

Figure 7-2 The change of the estrous cycle in different stages of the corpus luteum

统所包围（图 7-2，Ⅱ图），当黄体一分为二，裂点呈红色或褐色，而其余部分呈橙色或黄色；第Ⅲ阶段（发情周期第 11~17d）：当黄体上红色或褐色完全消失，呈鲜艳的橙色或黄色时第Ⅲ阶段开始，随后在这个阶段，脉管系统明显越过黄体裂点（图 7-2，Ⅲ图箭头）；在第Ⅳ阶段（发情周期第 18~20d），卵巢通常包含至少一个大的卵泡和一个退化的黄体（图 7-2，Ⅳ图箭头），且黄体表面无明显的脉管系统。

对于经产母牛，其发情周期内有 3 个卵泡波计，每个卵泡波开始的平均时间是发情周期的第 1.9d（范围 1~3d）、9.4d（8~11d）、16.1d

第七章
CART 及其候选受体基因在牛优势卵泡和从属卵泡中的表达

(14~19d);每个卵泡波中优势卵泡的最大直径分别为12.3~15.5mm、10.2~15.9mm、12.8~18.8mm;最大卵泡持续到发情周期的第6~6.4d、14.2~16d、21d。对于初情期母牛,在一个发情周期内只有2个卵泡波,第一和第二卵泡波出现于发情周期的第2~4d和10~11d,在第6d和19d分别达最大直径约13mm。因此,卵巢黄体不同阶段的形态特征结合母牛发情周期卵泡波存在的时间,可用来判断发情周期内的第一卵泡波,即发情周期开始的6d以内,此时黄体处于第Ⅰ~Ⅱ阶段之间。

在采样的过程中,获得的卵巢必须是非妊娠牛的2个卵巢;严格编号,保证2个卵巢属于同一头牛。黄体可以在屠宰场检查,也可以将一对卵巢放在一单独容器里实验室检查;检查每个卵巢的黄体,鉴别黄体所处阶段时仅需根据一个黄体的形态特征加以确定。检查黄体时要求寻找一5~6d的黄体,这样的黄体应当有排卵的裂点(被上皮细胞覆盖),黄体的顶点为红色或棕色(褐色)(图7-2,Ⅱ图)。

(二) 依据 E_2 和 P 水平确定第一卵泡波 DF 和 SF

卵泡发生闭锁的典型特征就是 E_2 的分泌能力丧失和卵泡内 E_2 与 P 比值的降低。E_2 分泌活跃的卵泡具有健康生长卵泡的生物化学特性和高的 E_2 : P 值;相对于闭锁卵泡或向闭锁趋势发育的卵泡,其 E_2 分泌能力低下,E_2 : P 值较低,其比值通常以 1 为分界。1995 年 Xu 等通过采集牛发情周期内出现第一卵泡波第 2d、4d、6d、8d、10d 的 DF,在第一卵泡波出现的第 6d,卵泡液 E_2 浓度开始降低,DF 中的 GCs,开始出现闭锁卵泡的形态学标志不容易检测,直到出现卵泡波的第 10d,DF 开始出现闭锁卵泡的形态学特征。表明在卵泡闭锁之前 DF 的 GCs 丧失分泌 E_2 的能力。此外,Austin 等也证明在牛发情周期的第一卵泡波中,在出现卵泡闭锁之前,SF 丧失了 E_2 分泌能力。有腔卵泡具有大量产生 E_2 的能力是健康状况卵泡的基本特征。因此,在本试验中采用第一卵泡波中最大卵泡卵泡液 E_2 : P > 1 和第二大卵泡卵泡液 E_2 : P < 1 来确定

DF 和 SF。

同时，在筛选卵泡的过程中还需注意：对于优势卵泡的筛选，要求该卵泡大小在 10~13mm，应是透明的或呈黄色，这样的卵泡才可能是卵巢上的 DF；不要使用红色的或大于 13mm 的卵泡，因为这样的卵泡可能是上一发情周期内卵泡波的退化卵泡；避免使用卵巢上存在多个 10~13mm 的透明卵泡，这种情况可能是共显性卵泡（Co–dominant follicles）；也不要使用存在 20mm 或更大卵泡的卵巢，这很可能是卵巢上出现囊肿，会影响优势卵泡的发育。SF 选择同一头牛的 2 个卵巢上第二大透明的或呈黄色的卵泡。在发情周期的这个阶段，第二大卵泡一般比 DF 小 1.5~2mm，同时，不要选择看起来呈红色或黑色的卵泡，这可能是上一发情周期卵泡波中的退化卵泡。

第二节 CART 及其候选受体在牛优势卵泡和从属卵泡的表达量分析

一、概述

qRT-PCR 是近年来广泛应用的基因表达量定量方法，具有准确率和灵敏度高的特点，在细胞生物学、分子生物学等领域得到广泛应用。该技术有特异性荧光探针法和非特异性荧光染料法之分，最常用的是 DNA 结合染料 SYBR Green Ⅰ 非特异性方法和 Taqman 水解探针特异性方法。SYBR Green Ⅰ 染料通过结合于双链 DNA 小沟区域与其结合后，荧光大大增强，在核酸定量检测方面具有很多优点：对模板选择范围广，应用程序通用性好，检测成本较低，灵敏度高。缺点是容易与非特异性双链 DNA 结合产生假阳性，在试验中可通过对熔解曲线进行分析，及时发现问题并对反应条件进行优化。

本试验通过荧光定量技术，对分子对接筛选出的 CART 候选受体基因及 CART 在牛发情周期内第一卵泡波 DF 和 SF 中的表达情况进行了检测，同时，也是对高通量测序结果可靠性的验证。

第七章
CART 及其候选受体基因在牛优势卵泡和从属卵泡中的表达

二、材料与方法

(一) 试验动物及其组织

本试验材料均来自山西省文水县胡兰镇贤美肉牛屠宰场。母牛屠宰后采集双侧卵巢上 2 个最大卵泡,按第一节的方法获得 DF 和 SF,并分别分离颗粒细胞。

(二) 主要试剂及耗材

Total RNA 提取试剂 Trizol、DEPC（Invitrogen 公司,美国）,固相 RNase 清除剂（Andybio 公司,美国）, PrimeScript© RT reagent Kit With gDNA Eraser 试剂盒、dNTP、DNA Marker DL2000（Takara,大连）,E_2 Elisa Kit 试剂盒、PROG Elisa Kit 试剂盒（DRG,德国）, PrimeScript© RT reagent Kit With gDNA Eraser、Rnase – free H_2O、SYBR© Premix Ex TaqTM Ⅱ、TaqDNA 聚合酶、2 × Es Taq Master Mix、dNTPs、$MgCl_2$（Takara,大连）,溴化乙锭（sigma,德国）,其余试剂为国产分析纯;引物合成由生工生物工程（上海）股份有限公司完成;液氮、无 RNase EP 管、八连管、无 RNase 枪头、PE 手套、橡胶手套、口罩、研钵、眼科剪和镊子等。

(三) 试验方法

1. 刮取颗粒细胞

卵泡一分为二剪开,但没有完全剪断。然后每一半卵泡平铺在表面皿上,使用一小刮刀（细胞刮棒）在卵泡内壁刮掉颗粒细胞（仅刮一次,避免刮下膜细胞）。然后将卵泡转移到一个加有培养液的有盖培养皿中进行隔离,准备一个 15ml 橙色、无菌、内置 2ml 培养液的带盖试管置于冰上,用细胞移液器将包含有颗粒细胞的培养液转移到试管中。表面皿上剩余的颗粒细胞用培养液清洗后转移到 15ml 的橙色试管中。

2. 颗粒细胞 Total RNA 的提取及完整性和浓度的检测

颗粒细胞 Total RNA 的提取及完整性和浓度的检测、反转录 PCR 反应过程参照第三章。

3. PCR 反应体系及反应条件

利用 Primer 3.0 在线设计软件，根据 NCBI 上 *Bos taurus*（*cattle*）的相关基因序列设计荧光定量引物；以牛 *RPLP0* 和 *GAPDH* 基因作为内参基因，参考 GenBank 上提供的序列设计引物，由生工生物工程（上海）股份有限公司合成。引物序列如表 7-2 所示。

表 7-2 荧光定量 PCR 所用引物

Table 7-2 Primers of quantitative PCR

基因类型	基因名称	引物
内参基因	*RPLP0*	F：caaccctgaagtgcttgacat R：aggcagatggatcagcca
	GAPDH	F：cctggagaaacctgccaagt R：agccgtattcattgtcatacca
目的基因	*CART*	F：cttatccttgcatccagacga R：ctagtaggctgcaaatccatgag
	TEDDM1	F：aatgtcccacagaggtttgc R：atgaatcttggtggcaggtc
	CMKLR1	F：acaccgtctggttcctcaac R：gttgctgatcttgcacatgg
	AGTR2	F：ttcccttccatgttctgacc R：aaggaagtgccaggtcaatg
	*GPR*116	F：tcgaggtgtcaacagtctgc R：taaggcaactgcagtgatgc

4. 荧光实时定量 PCR 反应体系与反应条件

扩增前需要进行标准曲线的制作，比较目的基因和内参基因扩增效率。依据标准曲线结果根据不同基因设置不同的扩增条件进行 qRT-PCR 反应，反应条件：95℃预变性 30s，95℃ 5s，60℃ 30s，72℃ 15s，45 个循环；95℃ 15s，60℃ 1min，95℃ 15s。目的基因和内参基因在同一条件下不同反应管内进行，反应结束后导出溶解曲线、动力学曲线和 Ct 值。

第七章
CART 及其候选受体基因在牛优势卵泡和从属卵泡中的表达

5. 统计分析

各目的基因的相对表达量采用△△CT 法计算,目的基因的相对表达水平 = $2^{-\triangle\triangle CT}$。用 Microsoft Excel 2007 对结果进行统计分析,qRT-PCR 相对表达量结果均用均值±标准差(Means±SD)表示,其中,各基因的表达量结果均应经内参基因 *RPLP*0 表达量的校正,全部数据采用 SPSS(Statistic package for social science 18.0)统计软件进行单因素方差分析检验。

三、结果与分析

(一)优势、从属卵泡颗粒细胞总 RNA 提取结果

分别对编号第 3、5、7 头牛卵巢 DF、SF 6 个样本进行总 RNA 的提取,样品经微量核酸蛋白测定仪纯度检测后,获得各样品总 RNA 的 OD260/OD280 均在 1.8~2.1,浓度均高于 1 000ng/μl,纯度和浓度均符合后续试验的要求。同时为了全面检测所提取颗粒细胞总 RNA 的质量,采用1%琼脂糖凝胶电泳检测,结果发现所提取各组总 RNA 的纯度比较高且完整性好,可直接用于下一步试验(图 7-3)。

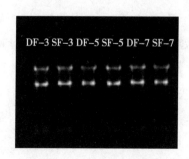

图 7-3 总 RNA 电泳图

Figure 7-3 Gel electrophoresis figure for total RNA

(二)CART 及其候选受体基因在第一卵泡波 DF 和 SF 中表达差异分析

根据 qRT-PCR 的技术要求,数据需通过 3 个方面的校正:一是重

复校正，要求最少 3 个以上的样本重复；二是物理校正，通过添加 ROX 荧光校正反应管之间的误差；三是内参基因校正，可校正因取样而产生误差。本实验通过预试验对 2 对内参基因进行效果验证，最终选择 *RPLP*0 作为内参基因，采用 3 个样本重复，2 个技术重复进行各基因的定量分析。

利用 △△CT 算法对 *CART*，*TEDDM*1，*CMKLR*1，*AGTR*2 和 *GPR*116 在牛发情周期内第一卵泡波 DF 和 SF 的表达量进行分析，统计结果如图 7 - 4 所示。结果显示：*CART* 和 *AGTR*2 在 DF 中的表达量极显著高于 SF（$P<0.01$）；*TEDDM*1 和 *CMKLR*1 在 SF 中的表达量显著高于 DF（$P<0.05$）；*GPR*116 在 DF 和 SF 的表达量不存在显著差异。

四、讨论

1995 年，美国 PE 公司成功推出的荧光定量 PCR 技术具有革命性意义，该技术成功融合了 PCR 技术的高灵敏性、DNA 杂交技术的高特异性和光谱技术的高精确定量等优点，通过直接探测 PCR 过程中荧光信号的变化，来获得特定区段扩增产物定量的结果。在 qRT – PCR 数据统计分析模型中，$2^{-\triangle\triangle CT}$ 法是广泛应用的统计方法，其前提是目的基因和持家基因的扩增效率一致，并通过模板的连续稀释建立标准曲线来测定扩增效率。李蘅（2008 年）采用 qRT – PCR 技术证实了 *Bcl* – 2、*Bax* mRNA 在山羊皮肤中存在差异表达，并根据 E_2 对其影响及所呈现的动态变化，揭示了在不同浓度 E_2 作用下 *Bcl* – 2、*Bax* 对皮肤毛囊生长的潜在作用。李拥军（2009 年）通过 qRT – PCR 检测了 *BMPl*5 mRNA 在山羊卵母细胞成熟发育不同阶段的表达情况，结果表明：在山羊卵母细胞成熟发育的不同阶段，*BMPl*5 表达量存在差异。在未成熟卵母细胞中该基因表达水平较低，培养 12h 后其表达量达到最大值并伴随有卵丘的扩展，随后下降，从而验证了 BMP15 在山羊卵母细胞成熟发育过程中发挥着重要的作用。陈雄云（2012 年）通过检测 *TYR*、*IGF* – 1 和

第七章
CART 及其候选受体基因在牛优势卵泡和从属卵泡中的表达

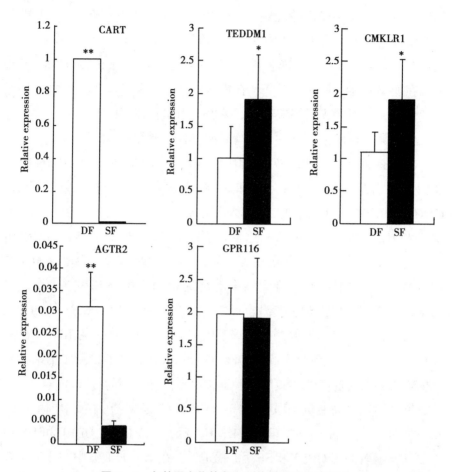

图 7-4 各基因在优势卵泡和从属卵泡中的表达

Figure 7-4 Real time PCR analysis of each gene expression in DF vs SF

注：上标的 ** 和 * 分别表示在显著水平 0.01 和 0.05 水平上的比较结果

Note: Superscript ** and * indicate significantly different at the level of 0.05 and 0.01.

FGF5 基因在不同毛色及同一毛色不同生长时期济宁青山羊皮肤的表达量，推测在济宁青山羊皮肤中，TYR 基因表达水平与毛色深浅呈正相关；验证了 IGF-1 基因表达的时空特性以及在济宁青山羊皮肤中 IGF-1 mRNA 的发育模式存在品种间差异；通过 FGF5 基因在济宁青山羊不同发育阶段皮肤差异表达的分析，推测 FGF5 基因对毛囊发育有抑

105

牛卵泡可卡因－苯丙胺调节转录肽（CART）受体的筛选

制作用。索峰（2012）通过 qRT – PCR 方法发现 $INH\beta_A$ 和 $INH\alpha$ 亚基基因在巴美肉羊发情后 1h、12h 和 24h 时间点，子宫和卵巢组织中相对表达量存在差异，在巴美肉羊发情后 1h、12h 和 24h，$INH\beta_A$ 和 $INH\alpha$ 亚基基因在卵巢的表达量先升后降，在排卵期抑制素在子宫和卵巢中的表达量变化较大且不同步，导致经产双羔巴美肉羊发情后比经产单羔巴美肉羊提前排卵，提前进入下一个发情周期，且排卵数量多于经产单羔羊。

本试验通过对 CART，TEDDM1，CMKLR1，AGTR2 和 GPR116 在牛发情周期内第一卵泡波中 DF 和 SF 的表达量进行分析，结果显示：CART 和 AGTR2 在 DF 中的表达量极显著高于 SF（$P < 0.01$），TEDDM1 和 CMKLR1 在 SF 中的表达量显著高于 DF（$P < 0.05$）；GPR116 在 DF 和 SF 的表达量不存在显著差异。其中可以看出，CART、AGTR2 和 GPR116 的 qRT – PCR 检测结果与高通量测序中该基因在 ODF1 和 ODF2 的 RPKM 的差异趋势基本一致，而 TEDDM1 和 CMKLR1 的 qRT – PCR 检测结果则与高通量测序中该基因在 ODF1 和 ODF2 的 RPKM 恰好相反，在 SF 中表达量显著高于 DF，这提示我们，高通量测序结果中的 RPKM 值必须通过其他实验加以验证，以提高其应用的可靠性。

同时，通过高通量测序和 qRT – PCR 检测发现，在牛发情周期内的第一卵泡波中，ODF1 和 ODF2、DF 和 SF 中，CART 在 ODF1 和 DF 中的表达量显著高于 ODF2 和 SF，也就是说，在第一卵泡波中的卵泡发育过程中当出现偏差后，CART 的表达量呈上升趋势。这提示我们卵泡发育过程中，在出现偏差之前大量卵泡发育走向凋亡，这个过程中可能存在其他较强的因素抑制了卵泡的发育；在出现偏差之后，随着卵泡的继续增大，CART 的表达量逐渐增加，从而抑制了 GCs 的增殖，造成 E_2 分泌量逐渐减少，最终导致 DF 走向闭锁而不能正常排卵。

神经肽首先要按照 mRNA 提供的模板，在胞体内经核糖体翻译成无活性的大分子神经肽前体。牛 CART 全长 116 个氨基酸，其中，$CART_{1-27}$ 为信号肽，前体在内质网切除信号肽形成神经肽原，向胞外转运过程中经多种酶的切割生成有生物活性的肽，本研究中进行分子对

第七章
CART 及其候选受体基因在牛优势卵泡和从属卵泡中的表达

接时所用 CART 活性肽段为 $CART_{76-116}$。通常，组织或细胞中 mRNA 与肽的含量有平行关系，因此，通过测定 mRNA 可反映组织或细胞中神经肽含量。但也有不少例外，组织或细胞中有 mRNA 的表达却不一定都能生成具有生物活性的肽。在试验中通过 qRT – PCR 证明 CART mRNA 在 DF 和 SF 中有表达且表达存在极显著差异，同时前期研究中，通过兔抗 CART 一抗对 CART 在牛卵泡中的表达研究也证明了 CART 在卵泡颗粒细胞层表达量较高，这说明在卵泡颗粒细胞层有 CART 活性肽的存在。

对于神经肽而言，效应细胞可能存在着多种受体，既可对同一递质或调质产生不同的反应，又可对不同的递质或调质产生不同的反应；也即同一种神经肽有多种功能，同一种神经肽对同一器官的效应也不一样，如内源性阿片肽除有镇痛效应外，还对循环、呼吸、消化、运动、内分泌、体温、免疫等功能起调节作用，造成机体内阿片肽多样化的生理功能。同时，神经肽也随剂量、作用部位的不同及动物种属的不同，肽的功能也不同；如低剂量黄体生成素释放激素（Luteinizing hormone releasing hormone，LHRH）合理使用有促进生育的功能，而高剂量长期连续使用有抑制生育的作用；α – MSH（α促黑激素）在两栖类和爬行类主要起调节皮肤颜色的作用，在哺乳类则可能起促进胎儿发育的作用。因此，神经肽的作用特点使得其受体的种类和作用特点更加复杂。在本试验中，*AGTR*2 在 DF 中的表达量极显著高于 SF（$P < 0.01$），*TEDDM*1 和 *CMKLR*1 在 SF 中的表达量显著高于 DF（$P < 0.05$）；通过 mRNA 表达量的变化也并不能排除其中一种蛋白非 CART 的受体。

神经肽受体的相对分子质量大多在 40 000 ~ 50 000Da，由 350 ~ 500 个氨基酸组成，肽链形成 7 个高度保守的跨膜区，N 端在细胞外，常有糖基的修饰，C 端在细胞内。受体肽链 C 端连接第 5 和第 6 个跨膜区段的胞内环是受体与 G 蛋白的结合部位。神经肽受体的识别部位是深埋在膜的内部，由 7 个跨膜区段间通过脯氨酸和甘氨酸残基之间的相互作用形成复杂的空间构象。GPR116 由 1 343 个氨基酸组成，相对分子质量为149 000Da，从神经肽受体特征上来说，GPR116作为CART的受体基本可以排除。

G蛋白偶联受体是一个庞大的跨膜蛋白受体家族，目前已经发现800多个成员，广泛参与感知、生殖、发育、生长、神经和精神等多种生命活动以及内分泌和代谢等多种生理过程；同时，与糖尿病、心脏病、肿瘤、免疫和感染性疾病、神经与精神疾病等重要疾病的发生、发展及治疗密切相关。基于G蛋白偶联受体在生理病理过程中的重要生物作用，这一蛋白家族也是目前最重要的药物作用靶标库，超过50%的临床用药物以及正在研发中的药物都作用于G蛋白偶联受体。因此，对CART受体展开研究，不仅为进一步分析单胎家畜卵泡闭锁提供理论依据，进而为提高单胎家畜的排卵率，解决优良种畜扩繁等提供技术支持；同时也为重大疾病基础研究以及新药研发奠定基础。

小结

1. 根据牛卵巢黄体的形态特征，从大量屠宰母牛群中获得11对处于发情周期第一卵泡波阶段的卵巢。

2. 应用ELISA技术对11对大、小卵泡卵泡液的E_2和P进行了测定，通过分别对大、小卵泡E_2:P值的计算，筛选出编号第3、5、7 3头牛的大小卵泡为DF和SF。

3. 对符合要求的3头牛卵巢DF、SF 6个样本进行总RNA的提取，经核酸蛋白测定仪纯度检测，OD260/OD280均在1.8~2.1，浓度均高于1 000ng/μl。

4. qRT-PCR检测结果显示：*CART*和*AGTR2*在DF中的表达量极显著高于SF（$P < 0.01$），*TEDDM*1和*CMKLR*1在SF中的表达量显著高于DF（$P < 0.05$）；*GPR*116在DF和SF的表达量不存在显著差异。

研究展望

　　CART 是一种下丘脑神经肽,在动物体内执行多种生理调节功能:包括抑制摄食、调节能量代谢平衡、调节机体应激反应、调控下丘脑—垂体—性腺轴和下丘脑—垂体—肾上腺轴、调节自主神经和感觉传导;同时,CART 与 Ⅱ 型糖尿病胰岛素抵抗有关;对动物卵泡发育有着显著的负调控作用,引起卵泡闭锁。因此,展开对 CART 受体的筛选及其功能研究,具有重要的现实意义和应用价值。

　　根据研究结果和存在问题,对 CART 受体的深入研究有以下设想。

　　第一,针对研究过程中存在的问题,后续研究中将对高通量测序结果进行进一步数据挖掘以获得一些有意义的基因,同时扩大差异表达基因筛选范围,并采用膜蛋白提取技术进行蛋白质组学研究。

　　第二,应用蛋白过表达技术和细胞结合试验明确 CART 的受体。

　　第三,选择合适的 CART 受体过表达系统,获得高纯度的 CART 受体蛋白溶液。

　　第四,应用 X 射线晶体衍射技术或核磁共振波谱技术获得 CART 及其受体蛋白复合体在溶液中的三维结构,并分析其动态变化过程。

　　第五,研究 CART 与相关蛋白的调控机制以及完善 CART 在颗粒细胞信号转导的分子作用机制;理清 CART 及其相互作用蛋白的网络图谱以及在复杂蛋白质体系的形成与功能发挥过程中关键作用。

第六，通过解析 CART 受体功能结构域的立体结构和分子组成，设计并开发针对性治疗 II 型糖尿病胰岛素抵抗的药物、靶向镇痛药物以及减肥药物。

目前，课题组继续深入研究，已经完成牛卵泡 CART 候选受体 AGTR2、TEDDM1 和 CMKLR1 全长序列的测定及其在颗粒细胞中的蛋白定量和定位研究，下一步将通过荧光标记技术、蛋白过表达技术及配体—受体结合试验，最终确定 CART 的受体，为进一步解析 CART 受体的立体结构及相关药物设计奠定基础。

附：本著作来源于以下已发表或待发表论文研究成果

1. Li Pengfei, Meng Jinzhu, Joe Folger, Huang Yang, Liu Yan, Yao Jianbo, George W Smith, Lv Lihua. Characterization of the bovine ovarian follicles at predeviation and onset of deviation Transcriptome [J]. Int J Data Min Bioin. (Under review).

2. Pengfei Li, Xiaolei Yao, Jinzhu Meng, JInyan Sun, zhu Meng, Jinyan Sun, Yang Huang, Yan Liu, Xiaolong Jiang, Jianwei Chen, Miaomaio Zhao, Junping He, Wenzhong Liu, Jianbo Yao, George W Smith, Lihua Lv. Study on CART Expression and its Relationship with Follicular Fluid Hormone Concentration in Pig Follicles of Different Sizes. Anim Reprod Sci. (Under review).

3. Pengfei Li, Xiaolei Yao, Jinzhu Meng, JInyan Sun, zhu Meng, Jinyan Sun, Yang Huang, Yan Liu, Xiaolong Jiang, Jianwei Chen, Miaomaio Zhao, Junping He, Wenzhong Liu, Jianbo Yao, George W Smith, Lihua Lv. Sequencing、Protein Locating and Expression of CART in ovary of hen. Anim Reprod Sci. (Under review).

4. 李鹏飞，吕丽华. 牛发情周期卵泡发育及其研究方法进展 [J]. 畜牧兽医学报, 2013, 44 (3)：333 - 339.

5. 李鹏飞，岳文斌，黄洋，孙晋艳，李晓明，庞钰莹，于雪静，贺俊平，孟金柱，任有蛇，吕丽华. *CART* 对绵羊卵巢卵泡颗粒细胞雌激素产生的影响 [J]. 畜牧兽医学报, 2013, 44 (6)：853 - 857.

6. 李鹏飞，岳文斌，庞钰莹，于雪静，黄洋，任有蛇，吕丽华. FSH和胰岛素对绵羊卵巢卵泡颗粒细胞体外培养的影响［J］. 畜牧兽医学报，2013，44（9）：1386－1391.

7. 李鹏飞，岳文斌，李富禄，黄洋，孙晋艳，朱芷葳，于秀菊，贺俊平，范瑞文，任有蛇，吕丽华. 可卡因—苯丙胺调控转录肽（CART）对猪卵巢卵泡颗粒细胞雌激素产生的影响［J］. 畜牧兽医学报，2012，43（12）：1879－1886.

8. 李鹏飞，孟金柱，谢建山，朱芷葳，刘岩，姜晓龙，陈建伟，姚晓磊，赵妙妙，吕丽华，基于牛ODF1与ODF2转录组卵泡发育候选基因挖掘及功能验证，畜牧兽医学报，2015，46（11）：1961－1966

9. 李鹏飞，孟金柱，刘岩，黄洋，陈建伟，姜晓龙，曹霞，姚晓磊，赵妙妙，吕丽华. 免疫共沉淀筛选牛卵泡CART相互作用蛋白的研究. 畜牧兽医学报，2015，46（12）：（发排）

10. 李鹏飞，孟金柱，于雪静，庞钰莹，杜海燕，李晓明，姚晓磊，赵妙妙，吕丽华. 外源性dsRNA对蛋鸡卵泡基因表达及发育的影响. 畜牧兽医学报，（审稿中）

11. 李鹏飞. 牛卵泡可卡因—苯丙胺调节转录肽（CART）受体的筛选［D］.太谷：山西农业大学博士学位论文. 2014.

12. Huang Y, Jin H, Chen J, Jiang X, Li P, Ren Y, Liu W, Yao J, Folger JK, Smith GW, Lv L. Effect of Vitamin D on basal and Luteinizing Hormone (LH) induced testosterone production and mitochondrial dehydrogenase activity in cultured Leydig cells from immature and mature rams ［J］. Anim Reprod Sci. 2015，158（7）：109－114.

13. Xiaolei Yao, Ruigao Song, Lei Shi, Pengfei Li, Yang Huang, Miaomiao Zhao, Lihua LV. Molecular mechanisms associated with Selenium regulation cell cycle progression by $p53/p21$ signal pathway to induce sheep spermatogonial stem cells proliferation and apoptosis in vitro［R］. Theriogenology. (Under review)

附：本著作来源于以下已发表或待发表论文研究成果

14. Y. Huang, X. L. Yao, J. Z. Meng, Y. Liu, X. L. Jiang, J. W. Chen, P. F. Li, Y. S. Ren, W. Z. Liu, J. B. Yao, J. K. Folger, G. W. Smith, L. H. Lv, Intrafollicular expression and potential regulatory role of Cocaine – and Amphetamine – Regulated Transcript(CART) in the ovine ovary. Domestic Animal Endocrinology 2015, 54（9）: 30 – 36.

15. Xiaolong Jiang, Xia Cao, Yang Huang, Jianwei Chen, Xiaolei Yao, Miaomiao Zhao, Yan Liu, Jinzhu Meng, Pengfei Li, Zhiyan Li, Jianbo Yao, George W Smith, Lihua Lv. Effects of treatment with Astragalus Membranaceus on function of rat leydig cells［J］. BMC Complementary and Alternative Medicine. 2015（15）: 261 – 264.

16. 李鹏飞, 李富禄, 于秀菊, 吕丽华. 猪下丘脑 CART 完整 CDS 区结构域分析［J］.山西农业大学学报（自然科学版）, 2011, 31（5）: 469 – 473.

17. 李鹏飞, 李富禄, 于秀菊, 吕丽华. 猪 CART mRNA 全 CDS 区序列的克隆与表达载体的构建［J］.福建农林大学学报（自然科学版）, 2011, 40（6）: 610 – 613.

18. 姚晓磊, 李鹏飞, 姜晓龙, 孟金柱, 黄洋, 刘岩, 陈建伟, 赵妙妙, 吕丽华. 卵泡发育相关基因在牛优势和从属卵泡颗粒细胞中表达的研究［J］.畜牧兽医学报, 2014, 45（12）: 1957 – 1963.

19. 姚晓磊, 宋瑞高, 吕丽华, 李鹏飞, 刘强, 黄洋, 陈建伟, 姜晓龙, 曹霞, 赵妙妙, 张贵花, 王永新, 石磊. 不同日龄犊牛睾丸中细胞周期调控基因 $CyclinB1$ 和 $p34^{cdc2}$ 表达的差异［J］.畜牧兽医学报, 2015, 46（2）: 245 – 250.

20. 孟金柱, 李鹏飞, 姜小龙, 陈建伟, 刘岩, 黄洋, 吕丽华. 蛋鸡卵巢 PRSS23 基因 CDS 序列分析［J］.基因组学与应用生物学, 2014, 33（1）: 77 – 81.

21. 朱芷葳, 贺俊平, 于秀菊, 李鹏飞, 程志学, 董常生, PAX3 转录因子在羊驼皮肤组织中的表达和定位分析［J］. 畜牧兽医学报,

2012, 43（5）: 729 - 734.

22. 朱芷葳, 贺俊平, 程志学, 王海东, 李鹏飞, 乔德瑞, 董常生. miRNA 在成年羊驼皮肤组织中的表达［J］. 畜牧兽医学报, 2010, 41（10）: 1342 - 1345.

23. 朱芷葳, 李拴英, 李慧锋, 李鹏飞, 许冬梅. miRNA 在皮肤组织中的研究进展［J］. 中国畜牧兽医, 2012, 7（39）: 77 - 79.

24. 赵妙妙, 姜晓龙, 陈建伟, 黄洋, 姚晓磊, 李鹏飞, 吕丽华. FSH 及 WNT 信号通路对绵羊卵泡颗粒细胞的增殖及雌激素分泌的影响［J］. 畜牧兽医学报, 2015,（已录用, 待出版）.

25. 黄洋, 孙晋艳, 李鹏飞, 靳辉, 吕丽华. 绵羊卵泡颗粒细胞体外培养条件的优化［J］. 山西农业科学, 2012, 40（2）: 161 - 163.

26. 庞钰莹, 岳文斌, 于雪静, 李鹏飞, 孟金柱, 刘岩, 李晓明, 黄洋, 任有蛇, 吕丽华. 绵羊卵巢卵泡颗粒细胞体外培养中促卵泡素和胰岛素浓度优化研究［J］. 山西农业大学学报（自然科学版）, 2012, 32（6）: 540 - 543.

27. 于雪静, 庞钰莹, 贺俊平, 李鹏飞, 姜晓龙, 吕丽华. 蛋鸡下丘脑 CART mRNA 序列测定及分析［J］. 畜牧兽医科技信息, 2013（4）: 21 - 23.

28. Zhu Z, He J, Jia X, Jiang J, Bai R, Yu X, Lv L, Fan R, He X, Geng J, You R, Dong Y, Qiao D, Lee KB, Smith GW, Dong C. MicroRNA - 25 functions in regulation of pigmentation by targeting the transcription factor MITF in alpaca (Lama pacos) skin melanocytes. Domestic Animal Endocrinology［J］. 2010, 38（3）: 200 - 209.

29. 朱芷葳, 贺俊平, 于秀菊, 程志学, 董常生. Mitf - M 在羊驼皮肤组织的表达与序列分析及免疫组织化学定位［J］. 中国农业科学, 2012, 45（4）: 794 - 800.

专利

申报发明专利一项（国家专利局已受理, 2015—09）.

参考文献
REFERENCES

[1] Roche J F, Mihm M, Diskin M, et al. A review of regulation of follicle growth in cattle [J]. J Anim Sci, 1998, 76 (2): 16-29.

[2] Ginther O J, Wiltbank M C, Fricke P M, et al. Selection of the dominant follicle in cattle [J]. Biol Reprod, 1996, 55 (4): 1 187-1 194.

[3] Ireland J J. Control of follicular growth and development [J]. Reprod Fertil Suppl, 1987, 34: 39-54.

[4] Ireland J J, Roche J F. Hypotheses regarding development of dominant follicles during a bovine estrous cycle [M]. The Netherlands Martinus Nijhoff Publishers, 1987.

[5] Spicer L J. Echternkamp S E. Ovarian follicular growth, function and turnover in cattle: A review [J]. J Anim Sci, 1986, 62 (4): 428-451.

[6] Wei X H, Qi L H, Chi X C. The effects of FSH on the phosphorylation of smad2/smad3 protein in rat ovarian granulosa cells. [J]. Acta Anatomica Sinica, 2007, 38 (2): 205-208.

[7] Pederson H G, Watson E D, Telfer E E. Analysis of atresia in equine follicles using histology, fresh granulosa cell morphology and detection of DNA fragmentation [J]. Reproduction, 2003, 125 (3): 417-423.

[8] Zeuner A, Muller K, Reguszynski K, et al. Apoptosis within bovine follic-

ular cells and its effect on oocyte development during *in vitro* maturation [J]. Theriogenology, 2003, 59 (5-6): 1 421-1 433.

[9] Wongsrikeao P, Kaneshige Y, Ooki R, et al. Effect of the removal of cumulus cells on the nuclear maturation, fertilization and development of porcine oocytes [J]. Reprod Domest Anim, 2005, 40 (2): 166-170.

[10] Yang M Y, Rajamahendran R. Morphological and biochemical identification of apoptosis in small, medium and large bovine follicles and the effect of follicle-stimulating hormone and insulin-like growth factor-I on spontaneous apoptosis in cultured bovine granulosa cells [J]. Biol Reprod, 2000, 62 (5): 1 209-1 217.

[11] Luciano A M, Modins S, Gandolfi F, et al. Effect of cell-to-cell contact on in vitro deoxyribonucleic acid synthesis and apoptosis responses of bovine granulosa cells to insulin-like growth factor-I and epidermal growth factors [J]. Biol Reprod, 2000, 63 (6): 1 580-1 585.

[12] Rajakoski E. The ovarian follicular system in sexually mature heifers with special reference to seasonal, cyclical, and left-right variations [J]. Acta Endocrinol, 1960, 34 (8): 57-68.

[13] Goodman A L, Nixon W E, Johnson D K, et al. Regulation of folliculogenesis in the cycling rhesus monkey: selection of the dominant follicle [J]. Endocrinology, 1977, 100 (1): 155-161.

[14] Ireland J J, Roche J F. Development of nonovulatory antral follicles in heifers: changes in steroids in follicular fluid and receptors for gonadotropins [J]. Endocrinology, 1983, 112 (1): 150-156.

[15] Ireland J J, Roche J F. Development of antral follicles in cattle after prostaglandin-induced luteolysis: changes in serum hormones, steroids in follicular fluid, and gonadotropin receptors [J]. Endocrinology, 1982, 111 (6): 2 077-2 086.

[16] Ireland J J, Roche J F. Growth and differentiation of large antral follicles after spontaneous luteolysis in heifers: changes in concentration of hor-

mones in follicular fluid and specific binding of gonadotropins to follicles [J]. J Anim Sci, 1983, 57 (1): 157-167.

[17] Dobson H. Plasma gonadotrophins and oestradiol during oestrus in the cow [J]. Reprod Fertil, 1978, 52 (1): 51-53.

[18] Adams G P, Matteri R I, Kastelic J P, et al. Association between surges of follicle - stimulating hormone and the emergence of follicular waves in heifers [J]. Reprod Fertil, 1992, 94 (1): 177-188.

[19] Miller K F, Critzer J K, Rowe R F, et al. Ovarian effects of bovine follicular fluid treatment in sheep and cattle [J]. Biol Reprod, 1979, 21 (3): 537-544.

[20] Good T E M, Ireland J L H, Weber P S D, et al. Isolation of nine different biologically and immunologically active molecular weight variants of bovine inhibin [J]. Biol Reprod, 1995, 53 (12): 1 478-1 488.

[21] Ireland J J, Curato A D, Wilson J. Effect of charcoal - treated bovine follicular fluid on secretion of LH and FSH in ovariectomized heifers [J]. Anim Sci, 1983, 57 (6): 1 512-1 516.

[22] Quirk S M., Fortune J E. Plasma concentrations of gonadotrophins, preovulatory follicular development and luteal function associated with bovine follicular fluid - induced delay of oestrus in heifers [J]. J Reprod Fertil, 1986, 76 (2): 609-621.

[23] Adams G P, Kot K, Smith C A, et al. Selection of a dominant follicle and suppression of follicular growth in heifers [J]. Anim Reprod Sci, 1993, 30 (4): 259-271.

[24] Mihm M, Good T E M, Ireland J L H, et al. Decline in serum follicle - stimulating hormone concentrations alters key intrafollicular growth factors involved in selection of the dominant follicle in heifers [J]. Biol Reprod, 1997. 57 (6): 1 328-1 337.

[25] Rahe C H, Owens R E, Fleeger J L, et al. Pattern of plasma luteinizing hormone in the cyclic cow: dependence upon the period of the cycle

[J]. Endocrinology, 1980, 107 (2): 498-503.

[26] Schallenberger E, Schondorfer A M, Walters D L. Gonadotrophins and ovarian steroids in cattle 1. Pulsatile changes of concentrations in the jugular vein throughout the oestrous cycle [J]. Acta Endocrinol, 1985, 108 (3): 312-321.

[27] Walters D L, Schams D, Schallenberger E. Pulsatile secretion of gonadotrophins, ovarian steroids and ovarian oxytocin during the luteal phase of the oestrous cycle in the cow [J]. J Reprod Fertil, 1984, 71 (7): 479-491.

[28] Cupp A S, Stumpf T T, Kojima F N, et al. Secretion of gonadotrophins change during the luteal phase of the bovine oestrous cycle in the absence of corresponding changes in progesterone or 17β-oestradiol [J]. Anim Reprod Sci, 1995, 37 (1): 109-119.

[29] Evans A C O, Komar C M, Wandji S A, et al. Changes in androgen secretion and luteinizing hormone pulse amplitude are associated with the recruitment and growth of ovarian follicles during the luteal phase of the bovine estrous cycle [J]. Biol Reprod, 1997, 57 (8): 394-401.

[30] Xu Z, Garverick H A, Smith G W, et al. Expression of follicle-stimulating hormone and luteinizing hormone receptor messenger ribonucleic acids in bovine follicles during the first follicular wave [J]. Biol Reprod, 1995, 53 (4): 951-957.

[31] Ireland J J, Mihm M, Austin E, et al. Historical perspective of turnover of dominant follicles during the bovine estrous cycle: Key concepts, studies, advancements, and terms [J]. J Dairy Sci, 2000, 83 (7): 1 648-1 658.

[32] Knight P G. Roles of inhibins, activins, and follistatin in the female reproductive system [J]. Front Neuroendocrinol, 1996, 17 (4): 476-509.

[33] Greenwald G S, Terranova P F. Follicular selection and its control [J].

Physiol Reprod, 1988, 2 (2): 387-445.

[34] Webb R R, Gosden G, Telfer E E, et al. Factors affecting folliculogenesis in ruminants [J]. Anim Sci, 1999, 68 (2): 257-284.

[35] Wei X H, Qi L H, Chi X C. The effects of FSH on the phosphorylation of smad2/smad3 protein in rat ovarian granulosa cells. [J]. Acta Anatomica Sinica, 2007, 38 (2): 205-208.

[36] Pederson H G, Watson E D, Telfer E E. Analysis of atresia in equine follicles using histology, fresh granulosa cell morphology and detection of DNA fragmentation [J]. Reproduction, 2003, 125 (3): 417-423.

[37] Bilodeau-Goeseels S, Panich P. Effects of oocyte quality on development and transcriptional activity in early bovine embryos [J]. Anim Reprod Sci, 2002, 71 (3-4): 143-155.

[38] Wongerikeao P, Kaneshige Y, Ooki R, et al. Effect of the removal of cumulus cells on the nuclear maturation, fertilization and development of porcine oocytes [J]. Reprod Domest Anim, 2005, 40 (2): 166-170.

[39] Yang M Y, Rajamahendran R. Morphological and biochemical identification of apoptosis in small, medium and large bovine follicles and the effect of follicle-stimulating hormone and insulin-like growth factor-I on spontaneous apoptosis in cultured bovine granulosa cells [J]. Biol Reprod, 2000, 62 (5): 1 209-1 217.

[40] Zeuner A, Muller K, Reguszynski K, et al. Apoptosis within bovine follicular cells and its effect on oocyte development during *in vitro* maturation [J]. Theriogenology, 2003, 59 (5-6): 1 421-1 433.

[41] Luciano A M, Modina S, Gandolfi F, et al. Effect of cell-to-cell contact on in vitro deoxyribonucleic acid synthesis and apoptosis responses of bovine granulosa cells to insulin-like growth factor-I and epidermal growth factors [J]. Biol Reprod, 2000, 63 (6): 1 580-1 585.

[42] Yao J, Ren X, Ireland J J, et al. Generation of a bovine oocyte cDNA library and microarray: resources for identification of genes important for

follicular development and early embryogenesis [J]. Physiological Genomics, 2004, 19 (1): 84 –92.

[43] Kobayashi Y, Jimenez – krassel F, Li Q, et al. Evidence that cocaine – and amphetamine – regulated transcript is a novel intraovarian regulator of follicular atresia [J]. Endocrinology 2004, 145 (11): 5 373 –5 383.

[44] Aritro S, Bettedowda A, Jimenez – krassel F, et al. Cocaine – and amphetamine – regulated transcript (CART) regulation of follicle stimulating hormone signal transduction in bovine granulosa cells [J]. Endocrinology, 2007, 148 (9): 4 400 –4 410.

[45] Lv L H, Jimenez – krassel F, Sen A, et al. Evidence supporting a role for cocaine and amphetamine regulated transcript (CART) in control of granulosa cell estradiol production associated with dominant follicle selection in cattle [J]. Biol Reprod, 2009, 81 (3): 580 –586.

[46] McNutt, W. G. The corpus luteum of the ox ovary in relation to the estrous cycle [J]. J Amer Vet Med Ass, 1924 (18): 556.

[47] Hammond J. The physiology of reproduction in the cow [M]. Cambridge University Press, London. 1927.

[48] Ireland J J, Murphee R L, Coulson P B. Accuracy of Predicting Stages of Bovine Estrous Cycle by Gross Appearance of the Corpus Luteum [J]. Journal of Dairy Science. 1980, 63 (1): 155 –160.

[49] Mitchison J M. The biology of the cell cycle [M]. Cambridge University Press, London. 1971.

[50] Fortune J E, Sirois J, Quirk S M. The growth and differentiation of ovarian follicles during the bovine estrous cycle [J]. Theriogenology, 1988, 29 (2): 95 –109.

[51] Savio J D, Keenan L, Boland M P, et al. Pattern of growth of dominant follicles during the oestrous cycle in heifers [J]. J Reprod Fertil, 1988, 83 (7): 663 –671.

[52] Sirois J, Fortune J E. Ovarian follicular dynamics during the estrous cycle

in heifers monitored by real – time ultrasonography [J]. Biol Reprod, 1988, 39 (2): 308 – 317.

[53] Erickson B H. Development and senescence of the postnatal bovine ovary [J]. Anim. Sci, 1966, 25 (3): 800 – 805.

[54] Marion G B, Gier H T, Choudary J B. Micromorphology of the bovine ovarian follicular system [J]. Anim. Sci, 1968, 27 (2): 451 – 465.

[55] Carson R S, Findlay J K, Clarke I J, et al. Estradiol, testosterone, and androstenedione in ovine follicular fluid during growth and atresia of ovarian follicles [J]. Biol Reprod, 1981, 24 (2): 105 – 113.

[56] Moor R M, Hay M F, Dott H M, et al. Macroscopic identification and steroidogenic function of atretic follicles in sheep [J]. J Endocrinol, 1978, 77 (6): 309 – 318.

[57] Roche J F. Control and regulation of folliculogenesis—a symposium in perspective [J]. Rev Reprod, 1996, 1 (1): 19 – 27.

[58] Webb R J G, Gong A, Law S, et al. Control of ovarian function in cattle [J]. J Reprod Fertil, (Suppl) 1992, 45: 141 – 156.

[59] Dufour J, Whitmore H L, Ginther O J, et al. Identification of the ovulating follicle by its size on different days of the estrous cycle in heifers [J]. Anim Sci, 1972, 34 (1): 85 – 87.

[60] Matton P, Adelakoun V, Couture Y, et al. Growth and replacement of the bovine ovarian follicles during the estrous cycle [J]. Anim Sci, 1981, 52 (3): 813 – 820.

[61] Choudary J B, Gier H T, Marion G B. Cyclic changes in bovine vesicular follicles [J]. Anim Sci, 1968, 27 (2): 468 – 471.

[62] Cupps P T, Laben R C, Mead S W. Histology of pituitary, adrenal, and reproductive organs in normal cattle and cattle with lowered reproductive efficiency [J]. Hilgardia, 1959 (29): 383 – 410.

[63] Ireland J J, Coulson P B, Murphree R L. Follicular development during the four stages of the estrous cycle of beef cattle [J]. Anim Sci, 1979,

49 (5): 1 261 - 1 269.

[64] Priedkalns J, Weber A F, Zemjanis R. Qualitative and quantitative morphological studies of the cells of the membrane granulosa, theca interna and corpus luteum of the bovine ovary [J]. Z Zellforsch, 1968, 85 (4): 501 - 520.

[65] Donaldson L, Hansel W. Cystic corpora lutea and normal and cystic graafian follicles in the cow [J]. Aust Vet J, 1968, 44 (7): 304 - 308.

[66] Reeves J J, Rantanen N W, Hauser M. Transrectal real - time ultrasound scanning of the cow reproductive tract [J]. Theriogenology, 1984, 21 (3): 485 - 494.

[67] Pierson R. A, Ginther O J. Ultrasonography of the bovine ovary [J]. Theriogenology, 1984, 21 (3): 495 - 504.

[68] Pierson R A, Ginther O J. Ultrasonic imaging of the ovaries and uterus in cattle [J]. Theriogenology, 1988, 29 (1): 21 - 37.

[69] Quirk S M, Hickey G J, Fortune J E. Growth and regression of ovarian follicles during the follicular phase of the oestrous cycle in heifers undergoing spontaneous and PGF - 2 - induced luteolysis [J]. J Reprod Fertil, 1986, 77 (1): 211 - 219.

[70] Sirois J, Fortune J E. Lengthening the bovine estrous cycle with low levels of exogenous progesterone: a model for studying ovarian follicular dominance [J]. Endocrinology, 1990, 127 (2): 916 - 925.

[71] Savio J D, Thatcher W W, Badinga L, et al. Regulation of dominant follicle turnover during the oestrous cycle in cows [J]. J Reprod Fertil, 1993, 97 (1): 197 - 203.

[72] Aritro S, Lv L H, Bello N, et al. Cocaine - and amphetamine - regulated transcript (CART) accelerates the temination of FSH induced ERK1/2 and AKT activation by regulating the expression and degradation of specific map kinase phosphatases [J]. Mol Endocrinol, 2008, 22 (12): 2 655 - 2 676.

[73] Wierup N, Kuhar M, Nilsson B O, et al. Cocaine – and amphetamine – regulated transcript (CART) is expressed in several islet cell types during rat development [J]. J Histochem Cytochem, 2004, 52 (2): 169 – 177.

[74] Guerardel A, Barat – Houari M, Vasseur F. et al. Analysis of sequence variability in the CART gene in relation to obesity in a Caucasian population [J]. BMC Genet, 2005, 6: 19.

[75] Douglass J, Mckinzie A A, Couccyro P. PCR differential display identifies a rat Brain mRNA that is transeriptionally regulated by cocaine and amphetamine [J]. The Journal of Neuroscience 1995, 15 (3Pt2): 2 471 – 2 481.

[76] Spiess J, Villarreal J, Vale W. Isolation and sequence analysis of a somatostatin – like Polypeptide from ovine Hypothalamus [J]. Bioehemistry. 1981; 20 (7): 1 982 – 1 988

[77] Ekblad E, Kuhar M, Wierup N, et al. Cocaine – and amphetamine – regulated transcript: distribution and function in rat gastrointestinal tract [J]. Neurogastroenterology & Motility, 2003, 15 (5): 545 – 557.

[78] Kristensen P, Judge M E, Thim L, et al. Hypothalamic CART is a new anorectic peptide regulated by leptin [J]. Nature, 1998, 393 (6680): 72 – 76.

[79] Stanley S A, Muprhy K G, Bewiek G A, et al. Regulation of rat Pituitary cocaine – and amphetamine – regulated transcript (CART) by CRH and glucocorticoids [J]. Am J Physiol Endocrinol Metab, 2004, 287 (3): 583 – 590.

[80] Smith S M, Vaughan J M, Donaldson C J, et al. Cocaine – and amphetamine – regulated transcript activates the hypothalamic – pituitary – adrenal axis through a corticotropin – releasing factor receptor – dependent mechanism [J]. Endocrinology, 2004, 145 (11): 5 202 – 5 209.

[81] Kuriyama G, Takekoshi S, Tojo K, et al. Cocaine – and amphetamine –

regulated transcript peptide in the rat anterior pituitary gland is localized in gonadotrophs and suppresses prolactin secretion [J]. Endocrinology, 2004, 145 (5): 2 542 – 2 550.

[82] Philip J L, Niels V, Poul C P, et al. Chronic Intracerebroventricular Administration of Recombinant CART (42 – 89) Peptide Inhibits Food Intake and Causes Weight Loss in Lean and Obese Zucker (fa/fa) Rats [J]. Obesity Research, 2000, 8 (8): 590 – 596.

[83] Koylu E O, Couceyro P R, Lambert P D, et al. Immunohistochemical localization of novel CART prptides in rat hypothalamus, pituitary, adreal gland [J]. Comp Neurol, 1998, 39 (1): 15 – 132.

[84] Broberger C. Hypothalamic cocaine – and amphetamine – regulated transcript (CART) neurons: histochemical ralationship to thyrotropin – releasing hormone, melanin – concentrating hormone, orexin/hypocretin and neuropeptide Y [J]. Brain Res, 1999, 848 (1 – 2): 101 – 113.

[85] Philpot K B, Dallvechia – Adams S, Smith Y, et al. A cocaine – and – amphetamine – regulated – transcript peptide projection from the lateral hypothalamus to the ventral tegmental area [J]. Neuroscience, 2005, 135 (3): 915 – 925.

[86] Hunter R G, Philpot K, Vicentic A, et al. CART in feeding and obesity [J]. Trends Endocrinol Metab, 2004, 15 (9): 454 – 459.

[87] 傅茂, 程桦, 严励, 等. 可卡因 – 苯丙胺调节转录肽基因与超重肥胖的关联研究 [J]. 中华内分泌代谢杂志, 2002, 18 (6): 435 – 438.

[88] Kobayashi Y, Jimenez – Krassel F, Ireland J J. Evidence of a local negative role for cocaine and amphetamine regulated transcript (CART), inhibins and low molecular weight insulin like growth factor binding proteins in regulation of granulose cell estradiol production during follicular waves in cattle [J]. Reproductive Biology and Endocrinology, 2006, 4 (1): 22 – 33.

[89] Lv L H, Jimenez - Krassel F, Sen A, et al. Evidence for Local Inhibitory Effect of the Cocaine - and Amphetamine - Regulated Transcript (CART) Peptide on LH - Induced Thecal Androstenedione Production [J]. Biology of Reproduction, 2007, 77: 179.

[90] Damaja M I, Zheng J F, Martina B R. Intrathecal CART (55 - 102) attenuates hyperlagesia and allodynia in a mouse model of neuropathic but not inflammatory pain [J]. Peptides, 2006, 27 (8): 2 019 - 2 023.

[91] Kuhar M J, Jaworski J N, Hubert G W, et al. Cocaine - and amphetamine - regulated transcript peptides play a role in drug abuse and are potential therapeutic targets [J]. Aaps J, 2005 (7): 259 - 265.

[92] Murphy K G. Dissecting the role of cocaine - and amphetamine - regulated transcript (CART) in the control of appetite [J]. Brief Funct Genomics Proteomics, 2005, 4 (2): 95 - 111.

[93] Broberger C, Cocaine - and amphetamine - regulated transcript (CART) and food intake: behavior in search of anatomy [J]. Drug Dev Res, 2000, 51 (2): 124 - 142.

[94] Aleksandra V, Anita L, Michael J K. CART (cocaine - and amphetamine - regulated transcript) peptide receptors: Specific binding in AtT20 cells [J]. European Journal of Pharmacology. 2005, 528 (1 - 3): 188 - 189.

[95] 陈娟. 可卡因安非他明调节转录肽相互作用蛋白的筛选 [D]. 上海: 第二军医大学博士学位论文, 2005.

[96] Douglass J, Daoud S. Characterization of the human cDNA and genomic DNA encoding CART: a cocaine - and amphetamine - regulated transcript [J]. Gene, 1996, 169 (2): 241 - 245.

[97] Gautvik K M, de Lecea L, Gautvik V T, et al. Overview of the most prevalent hypothalamus - specific mRNAs, as identified by directional tag PCR subtraction [J]. Proc Natl Acad Sci US A, 1996, 93 (16): 8 733 - 8 738.

[98] Vicentic A, Lakatos A, Kuhar M J. CART (cocaine - and amphetamine - regulated transcript) peptide receptors: specific binding in AtT20 cells [J]. Eur J Pharmacol, 2005, 528 (1-3): 188-189.

[99] Vicentic A, Lakatos A, Jones D, The CART receptors: background and recent advances [J]. Peptides, 2006, 27 (8): 1 934-1 937.

[100] Keller P A, Compan V, Bockaert J, et al. Characterization and localization of cocaine - and amphetamine - regulated transcript (CART) binding sites [J]. Peptides, 2006, 27 (6): 1 328-1 334.

[101] Iliff J J, Alkayed N J, Gloshani K J, et al. Cocaine - and amphetamine - regulated transcript (CART) peptide: a vasoactive role in the cerebral circulation [J]. J Cereb Blood Flow Metab, 2005, 25 (10): 1 376-1 385.

[102] Lakatos A, Prinster S, Vicentic A, et al. Cocaineand amphetamine - regulated transcript (CART) peptide activates the extracellular signal - regulated kinase (ERK) pathway in AtT20 cells via putative G - protein coupled receptors [J]. Neurosci Lett, 2005, 384 (1-2): 198-202.

[103] Yermolaieva O, Chen J, Couceyro P R, et al. Cocaine - and amphetamine - regulated transcript peptide modulation of voltage - gated Ca^{2+} signaling in hippocampal neurons [J]. J Neurosci, 2001, 21 (19): 7 474-7 480.

[104] Vrang N, Tang - Christensen M, Larsen P J, et al. Recombinant CART peptide induces c - Fos expression in central areas involved in control of feeding behaviour [J]. Brain Res, 1999, 818 (2): 499-509.

[105] Bannon A W, Seda J, Carmouche M, et al. Multiple behavioral effects of cocaine - and amphetamine - regulated transcript (CART) peptides in mice: CART 42 - 89 and ART 49 - 89 differ in potency and activity [J]. J Pharmacol Exp Ther, 2001, 99 (3): 1 021-1 026.

[106] Thim L, Kristensen P, Larsen P J, et al. CART, a new anorectic peptide [J]. Int J Biochem Cell Biol, 1998, 30 (12): 1 281-1 284.

[107] Dun S L, Brailoiu C G, Yang J, et al. Cocaine - and amphetamine - reg-

ulated transcript peptide and sympatho – adrenal axis[J]. Peptides, 2006, 27 (8): 1 949 – 1 955.

[108] Lenka M, Jana M, Resha M, et al. Cocaine – and amphetamine – regulated transcript (CART) peptide specific binding in pheochromocytoma cells PC12 [J]. European Journal of Pharmacology, 2007, 559 (2 – 3): 109 – 114.

[109] Smith D B, Johnson K S. Single – step purification of polypeptides expressed in Escherichia coli as fusions with glutathione S – transferase [J]. Gene, 1988, 67 (1): 31 – 40.

[110] Elands J, Barberis C, Jard S, et al. 125I – labelled d (CH2) 5 [Tyr (Me) 2, Thr4, Tyr – NH2 (9)] OVT: a selective oxytocin receptor ligand [J]. Eur J Pharmacol, 1988, 147 (2): 197 – 207.

[111] Koylu E O, Balkan B, Kuhar M J, et al. Cocaine and amphetamine regulated transcript (CART) and the stress response [J]. Peptides, 2006, 27 (8): 1 956 – 1 969.

[112] Venter J C, Adams M D, Myers E W, et al. The sequence of the human genome [J]. Science, 2001, 291 (5507): 1 304 – 1 351.

[113] Baltimore D. Our genome unveiled [J]. Nature, 2001, 409 (6822): 814 – 816.

[114] Tyers M, Mann M. From genomics to proteomics [J]. Nature, 2003, 422 (6928): 193 – 197.

[115] Tatusov R L, Koonin E V, Lipman D J. A genomic perspective on protein families [J]. Science, 1997, 278 (5338): 631 – 637.

[116] Swinbanks D. Government backs proteome proposal [J]. Nature, 1995, 378 (6558): 653.

[117] Wasinger V C, Cordwell S J, Cerpa – Poljak A, et al. Progress with gene – product mapping of the mollicutes: Mycoplasma genitalium [J]. Electrophoresis, 1995, 16 (7): 1 090 – 1 094.

[118] Margulies M, Egholm M, Altman W E, et al. Genome sequencing in

microfabricated high – density picolitre reactors [J]. Nature, 2005, 437 (7057): 376 – 380.

[119] Sanger F S, Nicklen A, Coulson R, et al. DNA sequencing with chain – terminating inhibitors [J]. Proc Natl Acad Sci USA, 1977, 74 (12): 5 463 – 5 467.

[120] Cohen A S, Najarian D R, Paulus A, et al. Rapid separation and purification of oligonucleotides by high – performance capillary gel electrophoresis [J]. Proc Natl Acad Sci USA, 1988, 85 (24): 9 660 – 9 663.

[121] Huang X C, Quesada M A, Mathies R A, et al. DNA sequencing using capillary array electrophoresis [J]. Anal Chem, 1992, 64 (18): 2 149 – 2 154.

[122] Bennett S T, Barnes C, Cox A, et al. Toward the 1 000 dollars human genome [J]. Pharmacogenomics, 2005, 6 (4): 373 – 382.

[123] Bentley D R. Whole – genome re – sequencing [J]. Curr Opin Genet Dev, 2006, 16 (6): 545 – 552.

[124] Audic S, Claverie J M. The significance of digital gene expression profiles [J]. Genome Res, 1997, 7 (10): 986 – 995.

[125] Golemis E. Protein – protein interactions: A molecular cloning manual [M]. Cold Spring Harbor, NY: Cold Spring Harbor Laboratory Press, 2002, 682.

[126] Masters S C. Co – immunoprecipitation from transfected cells [J]. Methods Mol Biol, 2004 (261): 337 – 350.

[127] Guttman M, Betts G, Barnes H, et al. Interactions of the NPXY microdomains of the low density lipop rotein receptor – related protein 1 [J]. Proteomics, 2009, 9 (22): 5 016 – 5 028.

[128] Li Z, Shi K, Guan L, et al. ROS leads to MnSOD upregulation through ERK2 translocation and p53 activation in selenite – induced apoptosis of NB4 cells [J]. FEBS Lett, 2010, 584 (11): 2 291 – 2 297.

[129] You Q H, Sun G Y, Wang N, et al. Interleukin – 17F – induced pulmo-

nary microvascular endothelial monolayer hyperpermeability via the protein kinase C pathway [J]. J Surg Res, 2010, 162 (1): 110 – 121.

[130] Issaq H J, Conrads T P, Janini G M, et al. Methods for fractionation, separation and profiling of proteins and peptides [J]. Electrophoresis. 2002, 23 (17): 3 048 – 3 061.

[131] Wilkins M R, Sanchez J C, Gooley A A. Progress with proteome projects: Why all proteins expressed by a genome should be identified and how to do it [J]. Biotech Gen Eng Rev, 1996, 13 (1): 19 – 50.

[132] Righetti P G, Castagna A. Recent trends in proteome analysis [J]. Adv Chromatogr, 2003, 42: 269 – 321.

[133] Fey S J, Larsen P M. 2D or not 2D. Two – dimensional gel electrophoresis [J]. Curr Opin Chem Biol, 2001, 5 (1): 26 – 33.

[134] Perkins D N, Pappin D J, Creasy D M, et al. Probability – based protein identification by searching sequence databases using mass spectrometry data [J]. Electrophoresis, 1999, 20 (18): 3 551 – 3 567.

[135] Geetha T, Langlais P, Luo M, et al. Label – Free Proteomic Identification of Endogenous, Insulin – Stimulated Interaction Partners of Insulin Receptor Substrate [J]. American Society for Mass Spectrometry, 2011 (22): 457 – 466.

[136] Hwang H, Bowen B P, Lefort N, et al. Proteomics analysis of human skeletal muscle reveals novel abnormalities in obesity and type 2 diabetes [J]. Diabetes, 2010, 59 (1): 33 – 42.

[137] Yu M, Wang J, Li W, et al. Proteomic screen defines the hepatocyte nuclear factor 1α – binding partners and identifies HMGB1 as a new cofactor of HNF1α [J]. Nucleic Acids Research, 2008, 36 (4): 1 209 – 1 219.

[138] Fenn J B, Mann M, Meng C K. Electrospray ionization for mass spectrometry of large biomolecules [J]. Science, 1989, 246 (4926): 64 – 71.

[139] Tanaka K, Waki H, Ido Y, et al. Protein and polymer analyses up to m/z 100000 by laser ionization time – of – fight mass spectrometry [J]. Rapid Commun Mass Spectrom, 1988, 2 (8): 151 – 153.

[140] Gevaert K, Vandekerckhove J. Protein identification methods in oteomics [J]. Electrophoresis, 2000, 21 (6): 1 145 – 1 154.

[141] Bakhtiar R, Nelson R W. Electrospray Ionization and Matrix – Assisted Laser Desorption Ionization Mass Spectrometry [J]. Biochemical Pharmacology, 2000, 59: 891 – 905.

[142] Aebersold R, Goodlett D R. Mass spectrometry in proteomics [J]. Chem Rev, 2001, 101 (2): 269 – 295.

[143] Page J S, Masselon C D, Smith R D. FTICR mass spectrometry for qualitative and quantitative bioanalyses [J]. Current opinion in Biotechnology, 2004, 15 (1): 3 – 11.

[144] Bakhtiar R, Nelson R W. Mass spectrometry of the proteome [J]. Mol Pharmacol, 2001, 60 (3): 405 – 415.

[145] Aebersold R, Mann M. Mass spectrometry – based proteomics [J]. Nature, 2003, 422 (6928): 198 – 207.

[146] Gershon D. Look, no hands [J]. Nature, 2003, 424 (6948): 583.

[147] Jain E. Current trends in bioinformatics [J]. Trends Biotechnol, 2002, 20 (8): 317 – 319.

[148] Stein L. Creating a bioinformatics nation [J]. Nature, 2002, 417 (6885): 119 – 120.

[149] 李玉岗. 生物大分子序列比对和蛋白质结构分类算法 [D]. 北京: 中国科学院研究生院博士学位论文, 2004.

[150] ZhouZ H, Towards atomic resolution structural determination by single – particle cryo – electron microscopy [J]. Curr Opin Struct Biol, 2008, 18 (2): 218 – 228.

[151] Lesk M A. Introduction to Protein Architecture [M]. Oxford UK: Oxford University Press, 2001.

[152] 卫艳丽. 基于蛋白质内源荧光磷光的分子识别作用研究 [D]. 太原: 山西大学博士学位论文, 2006.

[153] Chothia C, Lesk A M. The relation between the divergence of sequence and structure in proteins [J]. EMBO Journal, 1986, 5 (4): 823-836.

[154] Overignton J P. Comparison of three-dimensional structures of homologous proteins [J]. Curr Opin Struct Biol, 1992, 2 (3): 394-401.

[155] Swindells M B, Thornton J M. Modelling by homology [J]. Curr Opin Struct Biol, 1991, 1 (2): 219-223.

[156] Harvel T F, Snow M E. A new method for building protein conformations from sequence alignments with homologous of known structure [J]. J Mol Biol, 1991, 217 (1): 1-7.

[157] Lesk A M, Boswell D R. Homology modeling: inferences from tables of aligned sequences [J]. Curr Opin Struct Biol, 1992, 2 (2): 242-247.

[158] Lee R H. Protein model building using structural homology [J]. Nature, 1992, 356 (6369): 543-544.

[159] Harrison R W, Chatteijee D, Weber I T. Analysis of six protein structures predicted by comparative modeling techniques [J]. Proteins, 1995, 23 (4): 463-471.

[160] Cardozo T, Totrov M, Abagyan R. Homology modeling by the ICM method [J]. Proteins, 1995, 23 (3): 403-414.

[161] Mosimann S, Meleshko R, James N G. A critical assessment of comparative molecular modeling of tertiary structures of proteins [J]. Proteins, 1995, 23 (3): 301-317.

[162] Schrauber H, Eisenhaber F, Argos O. Rotamers, to be or not to be? [J]. J Mol Biol, 1993, 230 (2): 592-612.

[163] Laughton C A. Prediction of protein side-chain conformations from local three dimensional homology relationships [J]. J Mol Biol, 1994, 235 (3): 1 088-1 097.

[164] Jones T A, Thirup S. Using known substructures in protein model building and crystallography [J]. EMBO Journal, 1986, 5 (4): 819-823.

[165] Ripoll D R, Scheraga H A. On the multiple minima problem in the conformational analysis of polypeptides [J]. Biopolymers, 1990, 30 (1-2): 165-176.

[166] Summers N L, Karplus M. Modelling of side chains, loops and insertions in proteins [J]. Meth Enzyme, 1991, 202: 156-204.

[167] Jones D C, Kuhar M J. CART receptor binding in primary cell cultures of the rat nucleus accumbens [J]. Synapse, 2008, 62 (2): 122-127.

[168] Sen A, Lv L, Bello N, et al. Cocaine-and amphetamine-regulated transcript accelerates termination of follicle-stimulating hormone-induced extracellularly regulated kinase 1/2 and Akt activation by regulating the expression and degradation of specific mitogen-activated protein kinase phosphatases in bovine granulosa cells [J]. Mol Endocrinol, 2008, 22 (12): 2655-2676.

[169] Folger J K, Jimenez-Krassel F, Ireland J J, et al. Regulation of granulosa cell cocaine and amphetamine regulated transcript (CART) binding and effect of CART signaling inhibitor on granulosa cell estradiol production during dominant follicle selection in cattle [J]. Biol Reprod, 2013, 89 (6): 137-144.

[170] 王珑珑, 黄昊. G蛋白偶联受体的结构与功能——2012年诺贝尔化学奖相关研究成果简介 [J]. 生命科学, 2012, 24 (12): 1373-1379.

[171] Bentley D R, Balasubramanian S, Swerdlow H P, et al. Accurate whole human genome sequencing using reversible terminator chemistry [J]. Nature, 2008, 456 (7218): 53-59.

[172] Mardis E R. The impact of next-generation sequencing technology on genetics [J]. Trends Genet, 2008, 24 (3): 133-141.

[173] Marioni J C, Mason C E, Mane S M, et al. RNA - seq: an assessment of technical reproducibility and comparison with gene expression arrays [J]. Genome Res, 2008, 18 (9): 1 509 - 1 517.

[174] Croucher N J, Fookes M C, Perkins T T, et al. A simple method for directional transcriptome sequencing using Illumina technology [J]. Nucleic Acids Res, 2009, 37 (22): e148.

[175] Moore K, Thatcher W W. Major Advances Associated with Reproduction in Dairy Cattle [J]. Dairy Sci, 2006, 89 (4): 1 254 - 1 266.

[176] Carlos G G, Ralph J H, Evelyn E T, et al. Growth and antrum formation of bovine preantral follicles inlong - term culture in vitro [J]. Biol reprod, 2000, 62 (5): 1 322 - 1 332.

[177] Wilkins M R, sanehez J C, Gooley A A, et al. progress with proteome Projects: Why all proteins expressed by agenome should be identified and how to do it [J]. Bioteehnol Genet Eng Rev, 1996, 13 (1): 19 - 50.

[178] Klein J B, Thonghoonkerd V. Overview of proteomies [J]. Contrib NePhrol, 2004 (141): 1 - 10.

[179] Wasinger V C, Cordwell S J, Cerpapoljak A, et al. Progress with gene - product mapping of the mollieutes: Mycoplasma genitaliurn [J]. EleetroPhoresis, 1995, 16 (7): 1090 - 1094.

[180] Aderson N L, Anderson N G. Proteome and proteomics: new technologies, new concepts and new words [J]. Eleetrophoresis, 1998, 19 (11): 1 853 - 1 861.

[181] Lee K H. Proteomics: a techmology - driven and technology - limited discovery seience [J]. Trends Biotechool, 2001, 19 (6): 217 - 222.

[182] Cho W C, Proteomics technologies and challenges [J]. Geno. Prot. Bioinfo, 2007, 5 (2): 77 - 85.

[183] Patterson S D, Aebersold R H. Proteomics: the first decade and beyond [J]. Nat. Genet, 2003, 33 (Suppl): 311 - 323.

[184] Kowalski P, Stoerker J. Accelerating discoveries in the proteome and genome with MALDI TOF MS [J]. Phannacogenomies, 2000, 1 (3): 359–366.

[185] Cristoni S, Bemardi L R, Biunno I, et al. Analysis of protein ions in the range 3000~12000 Th under partial (no discharge) atmospheric pressure chemical ionization conditions using ion trap mass spectrometry [J]. Rapid Commun Mass Spectrom, 2002, 16 (12): 1 153–1 159.

[186] Grandori R. Electrospray–ionization mass spectrometry for protein conformational studies [J]. Curr Org Chem, 2003, 7 (15): 1 589–1 603.

[187] Rademaker G J, Thomas–Oates J. Analysis of glycoproteins and glycopeptides using fast–atom bombardment [M]. Methods in Molecular Biology: protein and peptide analysis by mass spectrometry, 1996 (61): 231–241.

[188] Aebersold R, Mann M. Mass spectrometry–based proteomies [J]. Nature, 2003, 422 (6928): 198–207.

[189] Cravatt B F, Simon G M, Yates J R. The biological impact of mass–spectrometry–based proteomics [J]. Nature, 2007, 450 (7172): 991–1 000.

[190] Demasi M A A, Montor W R., Ferreira G B, et al. Differential proteomic analysis of the anti–proliferative effect of glucocorticoid hormones in ST1 rat glioma cells [J]. Steroid Biochemistry and Molecular Biology, 2007, 103 (2): 137–148.

[191] Zhang L, Xie J, Wang X, et al. Proteomic analysis of mouse liver plasma membrane: use of differential extraction to enrich hydrophobic membrane proteins [J]. Proteomics, 2005, 5 (17): 4 510–4 524.

[192] Yu C S, Lin C J, Hwang J K. Predicting subcellular localization of proteins for Gram–negative bacteria by support vector machines based on n–peptide compositions [J]. Protein Science, 2004, 13 (5): 1 402–1 406.

[193] Yu C S, Chen Y C, Lu C H, et al. Prediction of protein subcellular localization [J]. Proteins: Structure, Function and Bioinformatics, 2006, 64 (3): 643 - 651.

[194] Huang D W, Sherman B T, Lempicki R A. Systematic and Integrative Analysis of Large Gene Lists Using DAVID Bioinformatics Resources [J]. Nat Protoc, 2009, 4 (1): 44 - 57.

[195] Huang D W, Brad T S, Richard A, et al. Bioinformatics enrichment tools: paths toward the comprehensive functional analysis of large gene lists [J]. Nucleic Acids Research, 2009, 37 (1): 1 - 13.

[196] Balgley B M, Laudeman T, Yang L, et al. Comparative evaluation of tandem MS search algorithms using a target - decoy search strategy [J]. Mol Cell Proteomics, 2007, 6 (9): 1 599 - 1 608.

[197] Yang X, Dondeti V, Dezube R, et al. DBP arser: web - based software for shotgun proteomic data analyses [J]. Proteome Res, 2004, 3 (5): 1 002 - 1 008.

[198] Nesvizhskii A I, Aebersold R. Interpretation of shotgun proteomic data: the protein inference problem [J]. Mol Cell Prot, 2005, 4 (10): 1 419 - 1 440.

[199] Stephan C, Reidegeld K A, Hamacher M, et al. Automated reprocessing pipeline for searching heterogeneous mass spectrometric data of the HUPO brain proteome project pilot phase [J]. Proteomics, 2006, 6 (18): 5 015 - 5 029.

[200] Feng J, Naiman D Q, Cooper B. Probability model for assessing proteins assembled from peptide sequences inferred from tandem mass spectrometry data [J]. Anal Chem, 2007, 79 (10): 3 901 - 3 911.

[201] Price T S, Lucitt M B, Wu W, et al. EBP, a program for protein identification using multiple tandem mass spectrometry datasets [J]. Mol Cell Proteomics, 2007, 6 (3): 527 - 536.

[202] Bader G D, Heilbut A, Andrens B, et al. Functional genomics and

protemic: Charting a multidimensional map of the yeast cell [J]. Trends Cell Biol, 2003, 13 (7): 344 –356.

[203] Moritz A L, Ji H, Schutz F, et al. A proteome strategy for fractionating proteins and peptides using for continuous free – flow electrophoresis coupled off line to reversed phase high performance liquid chromatography [J]. Anal chem, 2004, 76 (16): 4 811 –4 824.

[204] Patterson S D. Proteomics: the industalization of protein chemistry [J]. Curr opin Biotech, 2000, 11 (4): 413 –418.

[205] Jain K K. Proteomics: new technologies and their applications [J]. DDT, 2001, 6 (9): 457 –459.

[206] Walhout A J M, Vidal M. Protein interaction maps for model organisms [J]. Nature Reoiews Molecular Cell Biology, 2001, 2 (1): 55 –62.

[207] Chen Z, Han M. Building a protein interaction map: research in the post – genomic era [J]. Bioessays, 2000, 22 (6): 503 –506.

[208] Uetz P, Giot L, Cagney G, et al. A comprehensive analysis of protein – protein interactions in Saccharomyces cerevisiae [J]. Nature, 2000, 403 (6770): 623 –627.

[209] Schwikowski B, Uetz P, FieIds S. A network of protein – protein interactions in yeast [J]. Nature Biotechnology, 2000, 18 (12): 1 257 –1 261.

[210] Rain J C, Selig L, De Reuse H, et al. The protein – protein interaction map of Helicobacter pylori [J]. Nature, 2001, 409 (6718): 211 –215.

[211] Walhout A J, Sordella R, Lu X, et al. Protein interaction mapping in C. elegans using proteins involved in vulval development [J]. Science, 2000, 287 (5450): 116 –122.

[212] Wilm M, Shevchenko A, Houthaeve T, et al. Femtomole sequencing of proteins from polyacrylamide gels by nano – electrospray mass spectrometry [J]. Nature, 1996, 379 (6564): 466 –469.

参考文献

[213] Feng B B, Smith R D. A simple nanoelectrospray arrangement with controllable flowrate for mass analysis of submicroliter protein samples [J]. Journal of the American Society for Mass Spectrometry, 2000, 11 (1): 94 – 99.

[214] Zhou F, Shui W O, Lu Y, et al. High accuracy mass measurement of peptides with internal calibration using a dual electrospray ionization sprayer system for protein identification [J]. Rapid Commun Mass Spectrom, 2002, 16 (6): 505 – 511.

[215] Brownrs R S, Lennon J J. Mass resolution improvement by incorporation of pulsed ion extraction in a matrix – assisted laser desorption/ionization linear time – of – flight mass spectrometer [J]. Anal Chem, 1995, 67 (13): 1 998 – 2 003.

[216] Chait B T, Wang R, Beavis R C, et al. Protein ladder sequencing [J]. Science, 1993, 262 (5130): 89 – 92.

[217] Keough T, Youngguist R S, Lacey M P P. A method for high – sensitivity peptide sequencing using postsource decay matrix – assisted laser desorption ionization mass spectrometry [J]. Proc Natl Acad Sci USA, 1999, 96 (13): 7 131 – 7 136.

[218] Arnold R J, Karty J A. Observation of Escherichia coli ribosomal proteins and their posttranslational modifications by mass spectrometry [J]. Analytical Biochemistry, 1999, 269 (5): 105 – 112.

[219] Wilcox S K, Cavey G S, Pearson J D. Single ribosomal protein mutations in antibiotic resistant bacteria analyzed by mass spectrometry [J]. Antimicrobial Agents and Chemotherapy, 2001, 45 (11): 3 045 – 3 055.

[220] 其布勒哈斯. 应用 MALDI – TOF 质谱技术分析和鉴定食源性致病菌研究 [D]. 呼和浩特：内蒙古农业大学硕士学位论文, 2004.

[221] 尹晓. 肥胖症患者脂肪组织蛋白表达谱差异分析及 cathepsin K 功能初步研究 [D]. 上海：第二军医大学博士学位论文, 2001.

[222] 陈述林. 糖尿病人动脉血管差异表达蛋白质谱的建立和初步应用

[D]. 上海：上海第二医科大学博士学位论文，2004.

[223] 陈斌，毕志刚. 皮肤细胞蛋白质组质谱分析方法的建立及肽质量指纹图谱的构建 [J]. 中国中西医结合皮肤性病学杂志，2006，5 (3)：125 – 131.

[214] 解明然. 基于蛋白质组学技术筛选食管鳞癌相关蛋白的研究 [D]. 中山：中山大学博士学位论文，2010.

[225] 朱勇，沙家豪，周作明，等. 精子差异蛋白质谱分析父源性受精关键蛋白的研究 [C]. 中华医学会生殖医学分会人类精子库管理学组第三届年会暨全国男性生殖医学和精子库管理新进展第四次研讨会论文集，2012 – 09 – 06.

[226] Steen H, Mann M, The ABC´s（and XYZ´s）of peptide sequencing [J]. Nat Rev Mol Cell Biol，2004，5（9）：699 – 711.

[227] Leipzig J, Pevzner P, Heber S. The Alternative Splicing Gallery （ASG）：bridging the gap between genome and transcriptome [J]. Nucleic Acids Res，2004，32（13）：3 977 – 3 983.

[228] Frank A, Pevzner P. PepNovo：De novo peptide sequencing via probabilistic network modeling [J]. Anal Chem. 2005，77（4）：964 – 973.

[229] Deutsch E W, Lam H, Aebersold R. Data analysis and bioinformatics tools for tandem mass spectrometry in proteomics [J]. Physiol Genomics，2008，33（1）：18 – 25.

[230] 路长林. 神经肽基础与临床 [M]. 上海：第二军医大学出版社，2000.

[231] Tusnady G E, Simon I. Principles Governing Amino Acid Composition of Integral Membrane Proteins：Applications to Topology Prediction [J]. J Mol Biol，1998，283（2）：489 – 506.

[232] Tusnady G E, Simon I. The HMMTOP transmembrane topology prediction server [J]. Bioinformatics，2001，17（9）：849 – 850.

[233] Skouloubris S, Labihne A, Reuse H D. The AmiE aliphatic amidase and AmiF formamidase of Helicobacter pylori：natural evolution of two en-

zyme paralogues [J]. Mol Micro, 2001, 40 (3): 596 – 609.

[234] Browne W J, North A C T, Phillips D C, et al. A possible three – dimensional structure of bovine α – lactalbumin based on that of hen's egg – white lysozyme [J]. J Mol Biol, 1969, 42 (1): 65 – 70.

[235] Shotton D M, Watson H C. Three – dimensional structure of tosyl – elastase [J]. Nature, 1970, 225 (2): 811 – 816.

[236] Greer J. Model for haptoglobin heavy chain based upon structural homology [J]. Proc Nat Acad Sci USA, 1980, 77 (6): 3 393 – 3 397.

[237] Greer J. Comparative model – building of the mammalian serine proteases [J]. J Mol Biol, 1981, 153 (4): 1 027 – 1 042.

[238] Greer J. Model structure for the inflammatory protein C5a [J]. Science, 1985, 228 (4703): 1 055 – 1 060.

[239] Blundell T L, Sibanda B L, Sternberg M J E, et al. Knowledge – based prediction of protein structures and the design of novel molecules [J]. Nature, 1987, 326 (6111): 347 – 352.

[240] Blundell T, Carney D, Gardner S, et al. Knowledge – based protein modelling and design [J]. Eur J Biochem, 1988, 172 (3): 513 – 520.

[241] Altschul S F, Madden T L, Schaffer A A, et al. Gapped BLAST and PSI – BLAST: a new generation of protein database search programs [J]. Nucleic Acids Res, 1997, 25 (17): 3 389 – 3 402.

[242] Remmert M, Biegert A, Hauser A, et al. HHblits: lightning – fast iterative protein sequence searching by HMM – HMM alignment [J]. Nat Methods, 2012, 9 (2): 173 – 175.

[243] Guex N, Peitsch M C. SWISS – MODEL and the Swiss – PdbViewer: an environment for comparative protein modeling [J]. Electrophoresis, 1997, 18 (15): 2 714 – 2 723.

[244] Sali A, Blundell T L. Comparative protein modelling by satisfaction of spatial restraints [J]. J Mol Biol, 1993, 234 (3): 779 – 815.

[245] Benkert P, Biasini M, Schwede T. Toward the estimation of the absolute

quality of individual protein structure models [J]. Bioinformatics, 2011, 27 (3): 343 – 350.

[246] Mariani V, Kiefer F, Schmidt T, et al. Assessment of template based protein structure predictions in CASP9 [J]. Proteins, 2011, 79 (Suppl): 37 – 58.

[247] Arnold K, Bordoli L, Kopp J, et al. The SWISS – MODEL workspace: a web – based environment for protein structure homology modeling [J]. Bioinformatics, 2006, 22 (2): 195 – 201.

[248] Eswar N, Webb B, Marti – Renom M A, et al. Comparative Protein Structure Modeling With MODELLER [J]. Curr Protoc Bioinformatics, 2006, Chapter 5: Unit 5.6.

[249] Martí – Renom M A, Stuart A C, Fiser A, et al. Comparative protein structure modeling of genes and genomes [J]. Annu Rev Biophys Biomol Struct, 2000, 29 (1): 291 – 325.

[250] Fiser A, Do R K, Sali A. Modeling of loops in protein structures [J]. Protein Sci, 2000, 9 (9): 1 753 – 1 773.

[251] Jluy M, AmitA G, Alzari P M, et al. Crystal structure of a thermostable bacterial cellulose degrading enzyme [J]. Nature, 1992, 357 (6373): 89 – 91.

[252] Rouvinen J, Bergfors T, Teeri T. Three dimisional structure of cellobiohydrolase II from T. reesei [J]. Science, 1990, 249 (4967): 380 – 386.

[253] Sternberg M J E. Protein structure prediction – a practical approach [M]. Oxford: Oxford Press, 1996.

[254] Sanchez R, Sali A. Comparative protein structure modeling in genomics [J]. J Comput Phys, 1999, 151 (1): 388 – 401.

[255] Jiang T, XU Y, Zhang M Q. Current Topic in Computational Molecular Biology [M]. Bei Jing: Tinghua University Press, 2002.

[256] Pearson W R. Rapid and sensitive sequence comparison with FASTP and

FASTA [J]. Methods in Enzymology, 1990 (183): 63-98.

[257] Bacon D J, Anderson W F. Multiple sequence alignment [J]. J Mol Biol, 1986, 191 (2): 153-161.

[258] Schuler G D, Altschul S F, Lipman D J. A workbench for multiple alignment construction and analysis [J]. Proteins Struct Func Gen, 1991, 9 (3): 180-190.

[259] Berger M P, Munson P J. A novel randomized iterative strategy for aligning multiple protein sequences [J]. Comput Appl Biosci, 1991, 7 (4): 479-484.

[260] Wesson L, Eisenberg D. Atomic solvation parameters applied to molecular-dynamics of proteins in solution [J]. Protein Science, 1992, 1 (2): 227-235.

[261] Pascarella S A A P. Analysis of insertions deletions in protein structures [J]. J Mol Biol, 1992, 224 (2): 461-471.

[262] Allen F H, Kennard O. The cambridge crystallographic data centre: computer-based search, retrieval, analysis and display of information [J]. Chemical Design Automation News, 1979, B23: 2 331-2 339.

[263] Freyberg B Von, Richmond T J, Braun W. Surface-area included in energy refinement of proteins-a comparative-study on atomic solvation parameters [J]. J Mol Biol, 1993, 233 (2): 275-292.

[264] Macarthur M W, Thornton J M. Conformational analysis of protein structures derived from NMR data [J]. Proteins, 1993, 17 (3): 232-251.

[265] Dodge C. The HSSP database of protein structure sequence alignments and family profiles [J]. Nucleic Acid Res, 1998, 26 (1): 313-315.

[266] Pearson H. Virophage suggests viruses are alive [J]. Nature, 2008, 454 (7205): 677.

[267] Hummon A B, Richmond T A, Verleyen P, et al. From the genome to

the proteome: uncovering peptides in the Apis brain [J]. Science, 2006, 314 (5799): 647-649.

[268] Phizicky E, Bastiaens P I H, Zhu H, et al. Protein analysis on a proteomic scale [J]. Nature, 2003, 422 (6928): 208-215.

[269] Moreira I S, Fernandes P A, Ramos M J. Protein-protein docking dealing with the unknown [J]. J Comput Chem, 2010, 31 (2): 317-342.

[270] Ritchie D W. Recent progress and future directions in protein-protein docking [J]. Curr Protein Pept Sci, 2008, 9 (1): 1-15.

[271] Gray J J. High-resolution protein-protein docking [J]. Curr Opin Struct Biol, 2006, 16 (2): 183-193.

[272] Li C H, Ma X H, Chen W Z, et al. Progress in Protein-protein Docking Approaches [J]. Prog Biochem Biophys, 2006, 33 (7): 616-621.

[273] Piercel B, Tong W W, Weng Z P. M-ZDOCK: a grid-based approach for Cn symmetric multimer docking [J]. Structural bioinformatics, 2005, 21 (8): 1 472-1 478.

[274] Mintseris J, Pierce B, Wiehe K, et al. Integrating statistical pair potentials into protein complex prediction [J]. Proteins, 2007, 69 (3): 511-520.

[275] Chen R, Li L, Weng Z P. ZDOCK: An Initial-stage Protein Docking Algorithm [J]. Proteins, 2003, 52 (1): 80-87.

[276] Pierce B G, Hourai Y, Weng Z P. Accelerating Protein Docking in ZDOCK Using an Advanced 3D Convolution Library [J]. PLoS One, 2011, 6 (9): e24657.

[277] Pierce B G, Wiehe K, Hwang H, et al. ZDOCK server: interactive docking prediction of protein-protein complexes and symmetric multimers [J]. Bioinformatics, 2014, Epub ahead of print.

[278] Chen R, Weng Z: A novel shape complementarity scoring function for protein-protein docking [J]. Proteins, 2003, 51 (3): 397-408.

[279] Zhang C, Vasmatzis G, Cornette J, DeLisi C. Determination of atomic desolvation energies from the structures of crystallized proteins [J]. Mol Biol, 1997, 267 (3): 707-726.

[280] Hill S J. G-protein-coupled receptors: past, present and future [J]. Br J Pharmacol, 2006, 147 (S1): 27-37.

[281] Katritch V, Cherezov V, Stevens R C. Structure-function of the G protein-coupled receptor superfamily [J]. Annu Rev Pharmacol Toxicol, 2013, 53: 531-556.

[282] Rasmussen S G, Choi H J, Rosenbaum D M, et al. Crystal structure of the human β2-adrenergic G-protein-coupled receptor [J]. Nature, 2007, 450 (7168): 383-387.

[283] Rasmussen S G, DeVree B T, Zou Y, et al. Crystal structure of the β2 adrenergic receptor-Gs protein complex [J]. Nature, 2011, 477 (7366): 549-555.

[284] Park S H, Das B B, Casagrande F, et al. Structure of the chemokine receptor CXCR1 in phospholipid bilayers [J]. Nature, 2012, 491 (7426): 779-783.

[285] Ginestier C, Ginestier C, Diebel M E, et al. CXCR1 blockade selectively targets human breast cancer stem cells in vitro and in xenografts [J]. J Clin Invest, 2010, 120 (2): 485-497.

[286] Wu B, Chien Y E T, Mol C D, et al. Structures of the CXCR4 chemokine GPCR with small-molecule and cyclic peptide antagonists [J]. Science, 2010, 330 (6607): 1 066-1 071.

[287] Hanson M A, Stevens R C. Discovery of new GPCR biology: one receptor structure at a time [J]. Structure, 2009, 17 (1): 8-14.

[288] Zimin A V, Delcher A L, Florea L, et al. A whole-genome assembly of the domestic cow, Bos Taurus [J]. Genome Biol, 2009, 10 (4): 42-51.

[289] Yamazaki K, Adachi T, Sato K, et al. Identification and characteriza-

tion of novel and unknown mouse epididymis – specific genes by complementary DNA microarray technology [J]. Biol Reprod, 2006, 75 (3): 462 – 468.

[290] Gantz I, Konda Y, Yang Y K, et al. Molecular cloning of a novel receptor (CMKLR1) with homology to the chemotactic factor receptor [J]. Cytogenet Cell Genet, 1996, 74 (4): 286 – 290.

[291] Kaur J, Adya R, Tan B K, et al. Identification of chemerin receptor (ChemR23) in human endothelial cells: chemerin ~ induced endothelial angiogenesis [J]. Biochem Biophys Res Commun, 2010, 391 (4): 1 762 – 1 768.

[292] Arita M, Bianchini F, Aliberti J, et al. Stereochemical assignment, antiinflammatory properties, and receptor for the omega – 3 lipid mediator resolving EI [J]. J Exp Med, 2005, 201 (5): 713 – 722.

[293] Owman C, Nilsson C, Lolait S J. Cloning of cDNA encoding a putative chemoattractant receptor [J]. Genomic, 1996, 37 (2): 187 – 194.

[294] Meder W, Wendland M, Busmann A, et al. Characterization of human circulating TIG2 as a ligand for the orphan receptor ChemR23 [J]. FEBS Lett, 2003, 555 (3): 495 – 499.

[295] Wittamer V, Bondue B, Guillabert A, et al. Neutrophil ~ mediated maturation of chemerin: a link between innate and adaptive immunity [J]. J Immunol, 2005, 175 (1): 487 – 493.

[296] Zabel B A, Silverio A M, Butcher E C. Chemokine – like receptor 1 expression and chemerin – distinguish plasmacytoid from myeloid dendritic cells in human blood [J]. J Immunol, 2005, 174 (1): 244 – 251.

[297] Goralski K B, Mccarthy T C, Hanniman E A, et al. Chemerin, a novel adipokine that regulates adipogenesis and adipocyte metabolism [J]. J Biol Chem, 2007, 282 (38): 28 175 – 28 188.

[298] Takahashi M, Takahashi Y, Takahashi K, et al. Chemerin enhances insulin signaling and potentiates insulin – stimulated glucose uptake in

3T3～L1 adipocytes [J]. FEBS Lett, 2008, 582 (5): 573-578.

[299] Ichihara S, Senbonmatsu T, Price E Jr, et al. Targeted deletion of angiotensin II type 2 receptor caused cardiac rupture after acute myocardial infarction [J]. Circulation, 2002, 106 (17): 2 244-2 249.

[300] Oishi Y, Ozono R, Yano Y, et al. Cardioprotective role of AT2 receptor in postinfarction left ventricular remodeling [J]. Hypertension, 2003, 41 (3 Pt 2): 814-818.

[301] Virginie S V, Michael A B, Penny S E, et al. AGTR2 mutations in X-linked mental retardation [J]. Science, 2002, 296 (5577): 2 401-2 403.

[302] Rakugi H, Okamura A, Kamide K, et al. Recognition of tissue-and subtype-specific modulation of angiotensin II receptors using antibodies against AT1 and AT2 receptors [J]. Hypertens Res, 1997, 20 (1): 51-55.

[303] 王彤. 血管紧张素 II 及其受体在哮喘气道重塑中的作用及机制探讨 [D]. 南京: 南京医科大学博士学位论文, 2008.

[304] Fnkuzawa T, Hirose S. Multiple processing of Ig-Hepta/GPRl16, a G protein-coupled receptor with immunoglobulin (Ig) -like repeats, and generation of EGF2-like fragment [J]. J Biochem, 2006, 140 (3): 445-452.

[305] Tang X, Jin R, Qu G, et al. GPR116, an adhesion G-protein-coupled receptor, promotes breast cancer metastasis via the Gαq-p63 Rho GEF-Rho GTPase pathway [J]. Cancer Res, 2013, 73 (20): 6 206-6 218.

[306] Muller P Y, Janovjak H, Miserez A R, et al. Processing of gene expression data generated by quantitative real-time RT-PCR [J]. Biotechniques, 2002, 32 (6): 1 372.

[307] Bustin S A, Benes V, Nolan T, et al. Quantitative real-time RT-PCR a perspectiv [J]. J Mol Endocrionl, 2005, 34 (3): 597-601.

[308] Bustin S A. Real–time, fluorescence–based quantitative PCR: A snapshot of current procedures and preferences [J]. Expert Rev Mol Dlagn, 2005, 5 (4): 493–498.

[309] Schena L, Nigro F, Ippolito A, et al. Real–time quantitative PCR: A new technology to detect and study phytopathogenic and antagonistic fungi [J]. Eur J Plant Pathol, 2004, 110 (4): 893–908.

[310] Chen Y J, Hu C G, Zhao M Q. Construction of real–time quantitative polymerase chain reaction platform with SYBR Green I [J]. Prac J Med & Pharm, 2004, 21 (11): 997–999.

[311] Roche J F. Control and regulation of folliculogenesisa symposium in perspective [J]. Rev Reprod, 1996, 1 (1): 19–27.

[312] Sunderland S J, Crowe M A, Boland M P, et al. Selection, dominance and atresia of follicles during the oestrous cycle of heifers [J]. J Reprod Fertil, 1994, 101 (3): 547–555.

[313] Xu Z, Garverick H A, Smith G W, et al. Expression of messenger ribonucleic acid encoding cytochrome P450 side–chain cleavage, cytochrome p450 17 alpha–hydroxylase, and cytochrome P450 aromatase in bovine follicles during the first follicular wave [J]. Endocrinology, 1995, 136 (3): 981–989.

[314] Austin E J, Mihm M, Evans A C, et al. Alterations in intrafollicular regulatory factors and apoptosis during selection of follicles in the first follicular wave of the bovine estrous cycle [J]. Biol Reprod, 2001, 64 (3): 839–848.

[315] Pfaffl M W. A new mathematical model for relative quantification in real–time RT–PCR [J]. Nucleic Acids Res, 2001, 29 (9): e45.

[316] 李蘅. 绒山羊毛囊体外培养及雌二醇对毛囊 Bcl–2、Bax 表达的影响 [D]. 呼和浩特：内蒙古农业大学博士学位论文, 2008.

[317] 尚利青. EGCG 对山羊卵母细胞 IVM, IVF 的影响以及 BMP15 基因在山羊卵母细胞成熟过程的表达 [D]. 扬州：扬州大学硕士学位论

文，2009.

[318] 陈雄云. 山羊毛囊发育和毛色相关雉因的克隆与表达分析 [D]. 扬州：山东农业大学硕士学位论文，2012.

[319] 索峰. 抑制素 α 和 βA 亚基基因对巴美肉羊繁殖力影响的研究 [D]. 呼和浩特：内蒙古农业大学硕士学位论文，2012.

[320] 张伟，方泉，牟凌云，等. 以 G 蛋白偶联受体为靶点的神经肽创新药物研究 [A]. 第八届全国化学生物学学术会议论文摘要集，2013-09-15：90.

[321] Gaetano J S, Shawn R B, Carolina C, et al. Poster 8 hondroprotective effect of alpha-2-macroglobulin (A2M) on bovine cartilage explants [J]. PM&R, 2012, 4 (10): S191.

[322] Clarke E J, Allan V. Intermediate filaments: vimentin moves in [J]. Curr Biol, 2002, 12 (17): 596-598.

[323] Arun S, Shulin L. Vimentin in cancer and its potential as a molecular target for cancer therapy [J]. Cellular and Molecular Life Sciences, 2011, 68 (18): 3 033-3 046.

[324] Liu F, Lo C F, NING X, et al. Expression of NALP1 in cerebellar granule neurons stimulates apoptosis [J]. Cell Signal, 2004, 16 (9): 1 013-1 021.

[325] Lo C F, Ning X P, Gonzales C, et al. Induced expression of death domain genes NALP1 and NALP5 following neuronal injury [J]. *Biochem Biophys Res Commun*, 2008, 366 (3): 664-669.

[326] Christiansen H E, Schwarze U, Pyott S M, et al. Homozygosity for a missense mutation in SERPINH1, which encodes the collagen chaperone protein HSP47, results in severe recessive osteogenesis imperfect [J]. Am J Hum Genet, 2010, 86 (3): 389-398.

[327] Widmer C, Gebauer J M, Brunstein E, et al. Molecular basis for the action of the collagen-specific chaperone Hsp47/SERPINH1 and its structure-specific client recognition [J]. Proc Natl Acad Sci USA,

2012, 109 (33): 13 243 - 13 247.

[328] Eyre D R, Weis M A. Bone collagen: new clues to its mineralization mechanism from recessive osteogenesis imperfecta [J]. Calcified Tissue International, 2013, 93 (4): 338 - 347.

[329] Yu Y Y, Zeng B, Chail M, et al. Effect of serpinH1 on proliferation and growth of murine tumor [J]. Cancer Research on Prevention and Treatment, 2009, 36 (9): 741 - 744.

[330] Tufo G, Jones A W, Wang Z, et al. The protein disulfide isomerases PDIA4 and PDIA6 mediate resistance to cisplatin - induced cell death in lung adenocarcinoma [J]. Cell Death Differ, 2014, 21 (5): 685 - 695.

[331] Groenendyk J, Peng Z, Dudek E, et al. Interplay between the oxidoreductase PDIA6 and microRNA - 322 controls the response to disrupted endoplasmic reticulum calcium homeostasis [J]. Sci Signal, 2014, 7 (329): ra54.

[332] Vekich J A, Belmont P J, Thuerauf D J, et al. Protein disulfide isomerase - associated 6 is an ATF6 - inducible ER stress response protein that protects cardiac myocytes from ischemia/reperfusion - mediated cell death [J]. J Mol Cell Cardiol, 2012, 53 (2): 259 - 267.